Concepts of Optical Amplifiers

Concepts of Optical Amplifiers

Dr R.P. Mishra

Director
CSM College of
Engineering & Technology
Allahaba (U.P.)

SHREE PUBLISHERS & DISTRIBUTORS
NEW DELHI-110 002

Edition : 2011

ISBN : 978-81–8329–358–7

Published by :
SHREE PUBLISHERS & DISTRIBUTORS
22/4735 Prakash Deep Building,
Ansari Road, Darya Ganj,
New Delhi–110 002

Printed in India

Preface

The advent of optical communication as a primary source of information exchange is one of the biggest events of recent years. As the growth in internet users are extending exponentially, the bandwidth requirement and data speed are forcing the optical technology to make amazing advances. In that amplification of the signal in optical domain could not make pace with the requirements as electro-optic and all optical intergration require the compact devices on the pattern of VLSI. I have attmepted to encompass in this book all the optical amplifiers ranging from macro to micro in dimension to give wide choice to design engineers.

This Book is divided into Six Chapters. First Chapter "Physics of Optical Amplifiers" deals with the basic principles of optical amplification, especially the stimulated emission of Radiation principle. Brillouin and Raman Amplifiers use the physics of stokes waves to achieve downward or upward frequency shifts and amplification as well. Rare earth ions doped into fiber exhibit stark-splitting which is explained with the help of energy level diagram. Erbium doped fiber amplification process is explained in particular. Technology to enhance the amplification through use of optical cavity is explained.

The Second Chapter "Optical Amplifiers in Optical Communication" is the survey of all the existing technologies of optical amplifiers in long distance communication. A classification tree is presented and each type of amplifier is discussed to give insight into its design principle. Comparison table is presented based on different parameters and characteristics. Here in this chapter, it is realised that there is a need to design specific optical amplifiers to cover metro and access network as the optical fiber cable is expected to reach every home in the television sets and personal computers.

Third chapter deals with "Noise and Disturbances in Optical Amplification" that may degrade signal to Noise Ratio. Mathematical modelling of optical noise is presented. Interference, cross saturation and polarisation dependent gain problem is discussed and methods have been explored to reduce it which may help designing of new amplifiers.

Fourth Chapter is the Design of Optical Amplifiers for Metro and Access Application. A study of the issue involved in metro and access networks has been done, then in network configuration a small size, low cost specific optical amplifiers nick named as Amplets have been explored and their design and technology is explained to suit the need.

Fifth Chapter "Integrated Optical Amplifiers" focuses on the need of 40 Gb/S system. The integration of optical components or devices plays a key role in the design of high speed light wave system in terms of cost savings and complexity.

Chapter Six takes up "Fabrication of a Novel Integrated Amplifier : Optical Microring Amplifier". In this Chapter the fabrication of integrated optical waveguides and mircoring resonators with dopant like erbium is studied. It is found that sol-gel process is an effective method of fabricating microring amplifiers. In recent years there has been rapid progress in the fabrication and characterisation of sol-gel based IO waveguides and devices.

I expect the book is useful to the students, researchers and scientists working in the area of optical amplification. It is hoped that the approach adopted in the book will help the readers to acquire the new concepts and advanced methods and give him a sort of overall perspective necessary to enable him to continue with confidence further independent research.

It is fitting to record here my indebtness to all those learned people and authors of original papers who enlightened me. Throughout the book, I have attempted to give full references to all those authors whose work is quoted.

My gratitude to my research supervisor Prof. L. K. Singh, Director Institute of Physics and Electronics, Ex-Director Institute of Engineering and Technology, Dr. R.M.L. Avadh University, Faizabad for stimulating suggestions.

I am gratefully acknowledge Dr. T. Srinivas (Associate Professor, Indian Institute of Science) Prof. A. Selvarajan (Emeritus Professor, Indian Institute of Science), Dr. Vijayadithya Bh (Project Associate, Indian Institute of Science) for their valuable suggestions and creative inputs.

I have no words to thank my wife Dr. Pragya Mishra daughters Shambhavi and Ananya who missed me a lot during the busy days of my work but still co-operating and encouraging through out all the ups and downs of my life.

—Dr. R.P. Mishra

Contents

1

Physics of Optical Amplifiers

1.1 Introduction

In Optical Amplifiers, the light amplification is achieved just as in Lasers. In essence an optical amplifier is a laser with very low or without any optical feed back [1]. Semiconductor Optical Amplifier uses the principle of stimulated emission to amplify an optical information signal. Similarly fiber amplifiers also work on the principle of stimulated emission. In fiber amplifiers the energy delivered by a pump laser is used to excite atoms at the upper energy level, where they are stimulated by the photons of an information signal to fall to the lower level resulting in release of coherent light amplification [2].

1.2 Stimulated Absorption

Bohr's assumption was that electrons do not radiate or change their energy value during rotation. Any change in energy occurs only when electrons jump from one orbit to another. This implies that an entire atom possess discrete or quantized values of energy [3].

On the other hand interaction of light with atoms of the matter takes place in discrete packets of energy or quanta called photons such that absorption and emission of light causes them to make a transition from one energy state to another [4]. The frequency of the absorbed or emitted radiation 'f' is related to the difference in energy E between the higher energy state E_2 and the lower energy state E_1 by the expression.

$$E_{ph} = E_2 - E_1 = hf$$

Where h= 6.626×10^{-34} J.S. is Planck's Constant.

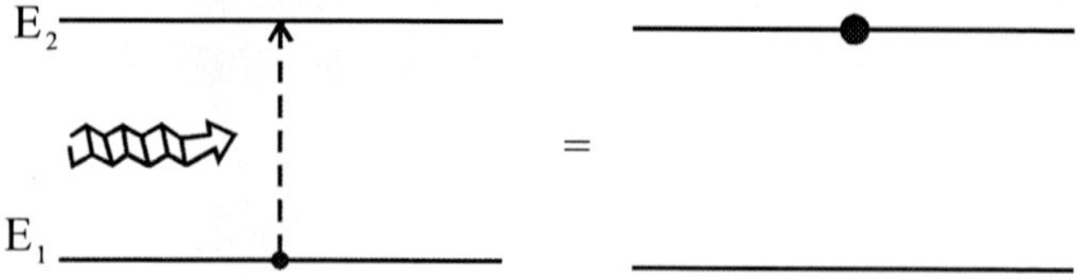

Fig 1.1 Two Energy Level Atomic System

This figure illustrates a two energy level atomic system where an atom is initially in the lower energy state E_1. When a photon with energy ($h_f = E_2 - E_1$) is incident on this atom then it absorbs the photon and attains the higher energy sate E_2. This is called stimulated absorption. When a photon with energy $E_{ph} < (E_2 - E_1)$ enters then it passes without interaction.

1.3 Emission of Radiation

Once the atom is excited in higher energy state then it can make a transition to the lower energy state E_1 in one of the following two ways.

(a)

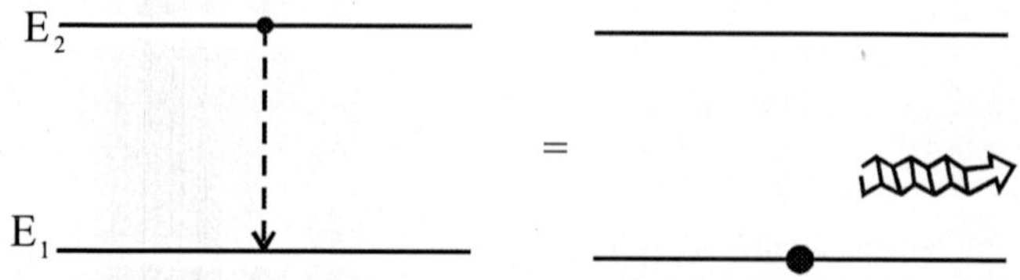

Fig1.2(a) Returning of Atom to Lower Energy state

The Figure (a) illustrate the returning of atom to the lower energy state in an entirely random manner making spontaneous emission. Random nature of the spontaneous emission process gives incoherent radiation.

The Figure (b) illustrate the interaction of photon $E_{ph} > E_2 - E_1$ with the atom in the upper energy state causing it to return to the lower state with the creation of a second Photon, making stimulated Emission.

(b)

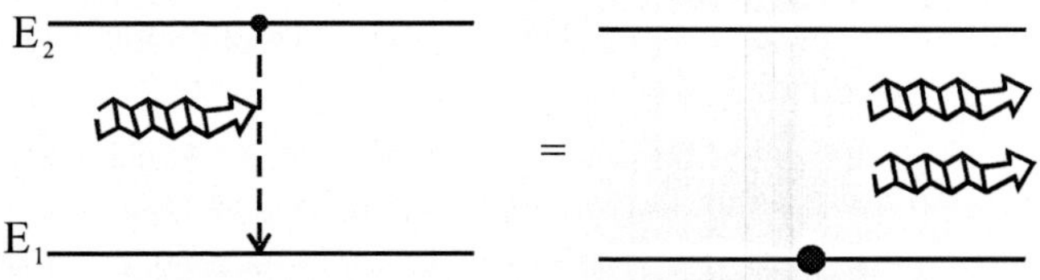

Fig 1.2(b) Returning of Lower Energy state after Interacting with Photon

1.4 Light Amplification

In stimulated emission, the light associated with stimulating and stimulated photon is in phase and has the same polarisation, therefore coherent radiation is obtained. In this coherent emission the liberated energy can add to the wave in a constructing manner called LASING to provide amplification [5].

1.5 Population Inversion

In order to produce sufficient amplification of light beam the rate of stimulated emission must be increased by making ($N_2>N_1$) i.e. population inversion. In certain materials the energy levels E_1 and E_2 have number of sublevels g_1 and g_2 respectively. This number of sublevels is called degeneracy parameter. If system is not degenerate then g_1 and g_2 may be set to unity. Due to Degeneracy nature, atoms have three or four energy levels which exhibit better population inversion than two level systems [6].

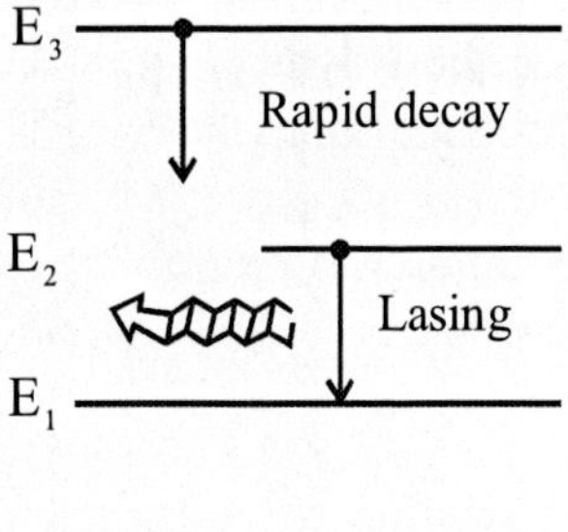

Fig 1.3(a) Three Level System

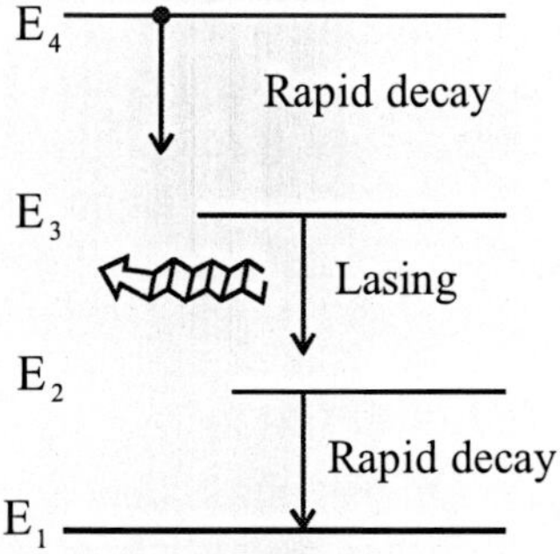

Fig 1.3(b) Four Level System

As shown in the above illustration, the central energy levels display central metastable state in which the atoms spend an unusually long time up to 10 m sec. It is from this metastable level that the stimulated emissions or lasing takes place. Metastable level exhibits a much longer life time which allows a large number of atoms to accumulate at this level. So over a period, after excitation or pumping, the density of atoms at metastable state increases above those in ground state and hence a population inversion is obtained between the two levels. Stimulated emission and hence lasing can then occur creating radiative electron transition.

A draw back with the three level system such as the ruby laser is that it generally requires very high pump power because the terminal state of the laser transition is the ground state. Hence more than Half the ground state atoms must be pumped into the metastable state to achieve population inversion. Where as a four level system such as the He-Ne laser is characterised by much lower pumping requirements. In this case the pumping excites the atoms from the ground state into energy level E_4 and they decay rapidly to metatable state E_3. However since the population of E_4 and E_2 remains essentially unchanged a small increase in the number of atoms in energy levels E_3 creates population inversion between E_3 and E_2 and lasing takes place between E_3 and E_2.

1.6 Travelling Wave

Population inversion ($N_2>N_1$) is achieved with the help of pumping So that now the incident wave is amplified by stimulated emission within the active area of laser. The confinement of energy is done to produce a very-very narrow beam. Further the amplification can be achieved in increased power terms, by making the active area longer so that further these two photons (one incident+ one stimulated emission) produce four photons. This avalanche effect continues till saturation. This process is called one way amplification and leads to travelling wave amplifiers. An optical amplifier which is not having feed back is called TWAs. [7]

1.7 Optical Cavity

Another approach other than travelling wave to achieve amplification is to have small area and then putting reflector at both the ends so that length is achieved by multiple reflections. Thus light amplification is achieved through optical cavity, also called as optical resonator. Semiconductor laser amplifiers are made by modifying the reflectivity of laser cavity. In SLA the facet reflectivity of resonant cavity is reduced to the level of 10^{-4} or even less [8]. This in turn will reduce the optical feedback.

1.8 Rare Earth Doped Fiber Amplifiers

Different Rare earth ions such as Erbium, Holmium, Neodymium, Thulium and Yetterbium can be doped to operate at different wave lengths. Erbium doped fiber amplifier have attracted the most attention among them simply because they operate near 1550 nm at which the fiber attenuation is minimum [10]. Fiber amplifiers have a particular advantage that they can be spliced into the system fiber with very low loss, and this is in contrast to large coupling loss associated with SLAs and fiber couplings. Therefore fiber amplifiers exhibit higher gain than SLAs.

1.9 EDFA Energy Level Diagram

When Erbium ions are incorporated through doping into a silica glass fiber, then each of the three energy levels of Erbium split into number of closely related multiple energy levels. This process is termed stark-splitting [9]. Stark splitting spreads each discrete energy level to a continuous energy band which is useful character of EDFA so that wavelength/frequency range of the signal that can be amplified is increased.

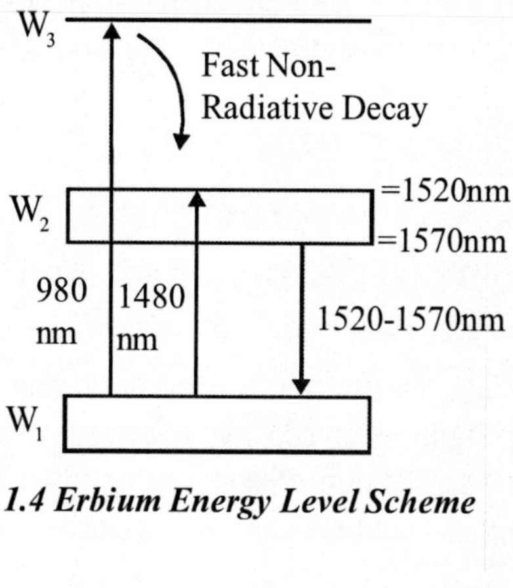

Fig 1.4 Erbium Energy Level Scheme

It's a gift of nature that the transition between W_2 and W_1 occurs at a set of wavelengths around 1550nm where silica fiber exhibits minimum attenuation.

Now to have more ions at level W_2, i.e for $N_2>N_1$, pumping is done. When 980 nm wavelength pumping is done, erbium ions are moved from W_1 to W_3 after just 1 msec, there they fastly and non-radiatively decay to W_2 where ions may stay upto 10 msec. Due to its long life time this state is called Metastable state. In this way such mechanism accumulates more ions at W_2, thus creating population inversion.

When 1480 nm pumping in done, ions are taken to W_2 state and thus accumulation starts here itself because only two levels take part thus achieving population inversion.

When an optical information signal now interacts while passing through such an inversely popuated Erbium Doped fiber, it stimulates the transition of erbium ions from level 2 to level 1. This stimulated transition will be accompanied by stimulated emission of photons having the same wavelength, direction, and phase of the input photon. Thus optical signals propagating through the EDFA with wavelength between about 1525 to 1565 nm induce stimulated emission in excited erbium ions and are there by amplified. In this way amplification. of input signal occur.

1.10 Erbium Doped Fiber Amplification

In 1990 Rare Earth doped fiber amplifiers emerged as a big competitor to semiconductor laser amplifiers. It is based on glass fibers doped with an appropriate amount of Erbium. EDFA consists of three basic components.

- Length of Suitable Erbium doped fiber
- Pump Laser
- Wavelength Selective Coupler

Pump laser can be introduced such that information signal and the pump signal co-propagate. (See fig a) An alternate geometry is possible where the pump laser can be placed so that info signal and pump signal counter propagate. (See fig b)

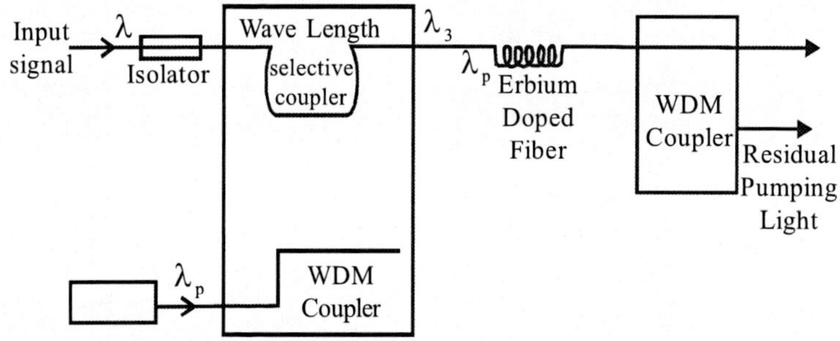

Fig. 1.5 Scheme of Erbium Doped Fiber Amplification

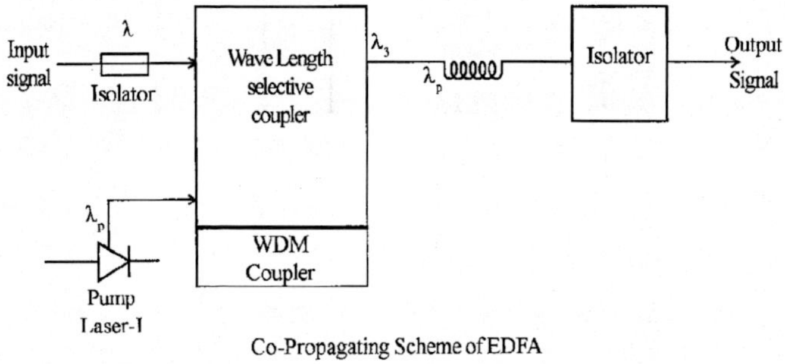

Fig. 1.6 (a) Co-Propagating Scheme of EDFA

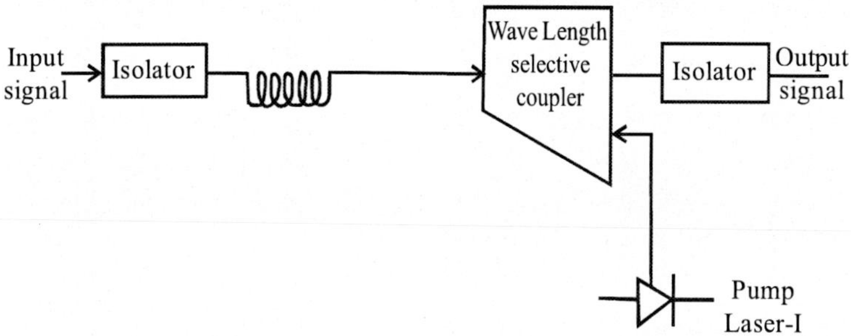

Fig. 1.6 (b) Counter Propagating Scheme of EDFA

The information signal is transmitted in the vicinity of 1550nm and pump laser radiate at 980nm or 1480nm. Here in the figure both the optical information and pumping beams are put in the same fiber by a WDM coupler.

The two beams propagate together along the doped section of the fiber, Where the information signal is amplified and pumping signal loses its power. In this way pumping light gives its power to an information signal and "dies". A co-propagating pump provides low noise and low power, whereas counter propagating pump provides higher output power and higher noise. So commercially bidirectional pumping is used [11].

Amplification in an EDFA occurs through the mechanism of stimulated emission. Energy from the pumping signal excites erbium ions at the upper energy band. The information signal then stimulates transition of the excited ions to the lower energy band. These transitions result in the radiation of photons with the same energy that is the same wavelength the input signal has. EDFA has relatively wide gain band width so it can amplify many wavelengths (Channels) simultaneously.

1.11 Fiber Raman Amplification

Fiber Raman amplifier works on the principle of stimulated raman scattering, which is based on Raman effect. In 1928 Sir CV Raman noticed the appearance of additional spectral lines after a monochromatic light passed through transparent material. It occurs owing to scattering effect in the fiber medium due to interactions of light waves with phonons (molecular vibrations) in silica medium [12].

During Raman Scattering, light incident on a medium is converted to a lower frequency. The spectral analysis of the scattered radiation reveals the existence of frequencies that are shifted down by increments equal to vibrational frequencies of molecules of medium. The physics of this phenomena can be explained as below.

When a monochromatic light beam gets scattered by a transparent substance then

(1) Over 99% of the scattered radiation has the same frequency as that of the incident light beam; this is known as Rayleigh scattering and it results from in homogeneities of a random nature. These inhomogeneities manifest to themselves as refractive index fluctuations and arise from density and

compositional variations which are frozen into the glass lattice on cooling [13]. The sky looks blue because of rayleigh scattering and the light that comes out from the sides of the optical fiber is also due to Rayleigh linear scattering.

(2) Around 1% of the scattered radiation has a frequency different from that of the incident beam because

(a) The incident beam may lead to translatory motion of the molecules which would result in producing a phonon (a quantum of an elastic wave) of acoustic frequency that produces an optical frequency shift which is usually very small so difficult to measure [13]. This is known as brillouin non-linear scattering.

(b) A part of the incident photon energy hn may be taken over by the molecule which may lead to rotational or vibrational motion. Which would result in producing phonons of high optical frequency. This phenomena leads to Raman-Stokes lines.

(c) On the other hand, there can be molecules which are already in excited state so when light is incident on such molecule the photons get scattered, and the molecule can de-excite to one of the lower energy states and in this process, some of the incident photon take up this excess energy and come out with a higher frequency, this leads to Raman anti-stokes lines. In this way quantity (hn-hn') or (hn'-hn) is usually referred to as the "Raman Shift" and is independent of the frequency of the incident radiation.

In stokes scattering, a spectral analysis of the scattered radiation reveals the existence of frequencies that are shifted down by increments equal to vibrational frequency of the materials.

Thus during raman scattering, the light incident on a medium is converted to a lower frequency. A pump photon Ep excites a molecule upto an excited state. From there the molecule quickly decays to a lower energy level emitting a signal photon Es in the process. The difference in energy between the pump and signal photons is dissipated as the molecular vibrations of the host material. For high pump powers, the scattered light can grow rapidly with most of the pump energy converted into scattered light giving rise to stimulated raman scattering which is the gain mechanism responsible for Raman amplification [14].

1.12 Brillouin Amplification

In stimulated Brillouin Scattering, the incident photon produces a phonon of acoustic frequency and as well as a scattered photon. This produces an optical frequency shift which varies with the scattering angle because the frequency of the sound wave varies with acoustic wavelength. The frequency shift is a maximum in the back ward direction reducing to zero in the forward direction making SBS a mainly backward process [15].

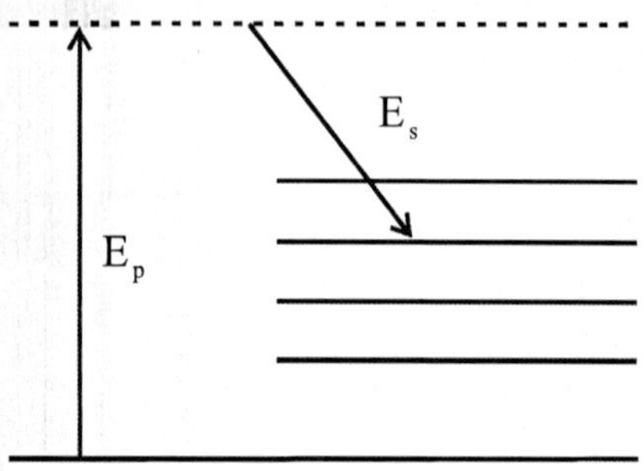

Thus SBS process results in the generation of the stokes wave which is down ward shifted in frequency from that of the incident wave. The interference between the pump and stokes waves tends to increase the amplitude of the acoustic, which in turn again interacts with the pump wave to reinforce the stokes wave. Under these conditions the process may become regenerative and cause exponential growth of the stokes wave leading to amplification.

1.13 Conclusion

In this Chapter different amplification principles are explained to understand the concept of amplification process. Technology to enhance the amplification through use of optical cavity is discussed. Stimulated emission is found to increase the effectiveness of amplification. Non-linear scattering phenomena is discussed to make use of its amplification attribute in multichannel transmission links.

References

[1] MJ O'mahony; "Semiconductor laser optical amplifiers for use in future fiber systems" JLT, 6(4), pp 531-544, 1988

[2] NA Olsson, "Light wave system with optical Amplifiers", JLT, Vol. 7, No. 7, pp 1071-1082, 1989

[3] C. Kittel, "Introduction to Solid state physics," seventh Edition, John wiley & Jons, pp 58-60, 2004

[4] CF Bohren and dR Huffman, "Absorption and Scattering of light by small particles"; John wiley & Sons, pp 157-165, 1998

[5] M. Hirao et all; "High reliability semiconductor lasers for optical communication"; IEEE Journal on selected areas in communication, 1986 pp 1494-1501

[6] John M Senior, "Optical Sources; The laser", Optical Fiber Communication<, PHI, 2004 pp. 288-295

[7] T. Nakagami and T Sakurai, "Optical and optoelectronic devices for optical fiber transmission system", IEEE commn magazine, 26(1), pp. 28-33, 1988

[8] Simon JC; "Ga In As P semiconductor laser amplifiers for Single mode fiber communications", IEEE, JLT-5 (1987), pp. 1286-1295

[9] Ravij Ramaswamy, Kumar N. Shivrajan; "Optical Networks"' Morgan Kaufman Publishers 2006, pp. 140-145

[10] Kasten Rottwit, et all, "Long distance transmission through distributed Erbium doped fibers"; JLT, Vol.-II, No. 12, Dec 1993

[11] DN ehen and E Desurvire, "Noise performance evaluation of distributed eribum doped fiber amplifiers with bidirectional pumping at 1.48 um"; IEEE Photon Tech. Lett, Vol.-4, No.1, pp. 52-55, 1992

[12] MN Islam; "Raman Amplifiers for telecommunication" Springer Publication, 2003, pp. 18-24

[13] J Shroeder, R Mohr et all, "Rayleigh and Brillouin Scattering in K_2O–SiO_2 glasses"; J. of American ceramic society, 56, pp 510-514, 1973

[14] Jake Bromage; "Raman Amplification for fiber communication systems"; JLT Vol. 22, No.-1, pp 79-93, 2004

[15] John M Senior; "Transmission Characteristics of optical fiber"; Optical Fiber Communication; Pearson education; pp. 94-96, 2007

2

Optical Amplifiers in Optical Communication

2.1 Introduction

Like any electronic communication system, the optical communication also involves a series of components. The main blocks of the system are transmitter and receiver, but for applications in long haul communication the installation of signal repeaters or line amplifiers become a necessity.

Like electronic communication the optical communication can be wireless or wire line (fiber). Therefore based on the type of channel (medium) used, the optical communication is classified into two categories -

(i) Optical Fiber Communication

(ii) Optical wireless Communication

The main blocks of optical communication system are shown in Fig 2.1.

Conventional repeaters employ opto electronic conversion. In which, the optical signal is first converted to an electrical signal and then amplified and reshaped electrical signal is converted again to optical domain for retransmission. This results in loss of speed and data rate[1]. Here if the optical signal could be directly amplified in the optical domain itself, the repeaters will be smaller

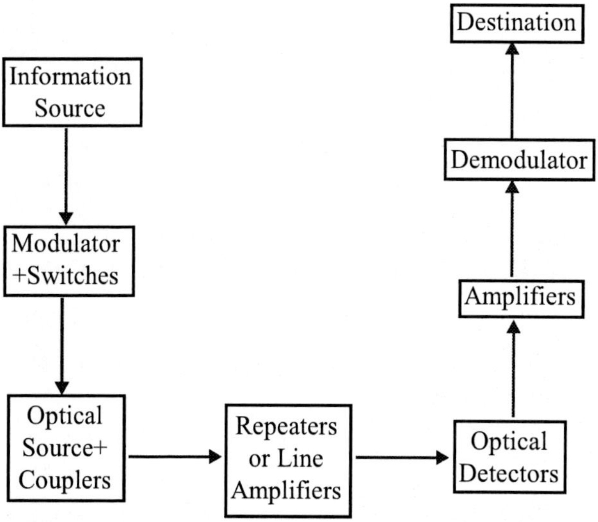

Fig 2.1 Main blocks of Optical Communication System

and cheaper which will enhance the speed of operation and avoid the signal distortions of electrical regime to notice that conversion of optical signal first to electrical signal and then reshaping it to required manipulations at any stage will again limit the benefits of optical communication to electrical communication. Therefore if optical communication could be done in optical domain then the benefits of optical band width, data rate and proper speed may be achieved.

There are two types of application for a linear amplifier in optical communication : one is a preamplifier that boosts a weak received signal in front of a photodetector and the other is a post amplifier that compensates for a transmission loss. Even if the intrinsic SNR of an optical signal is degraded by amplification still it will be better than, electronic signal processors accompanying large thermal noise, and therefore a noisy optical amplifier can still improve the overall system performance.[2]

The advent of optical amplifiers have brought a revolution in optical communication [3-6]. in last decade of 20th centuary and they have found their place not only as a repeater but also as

Classification Tree

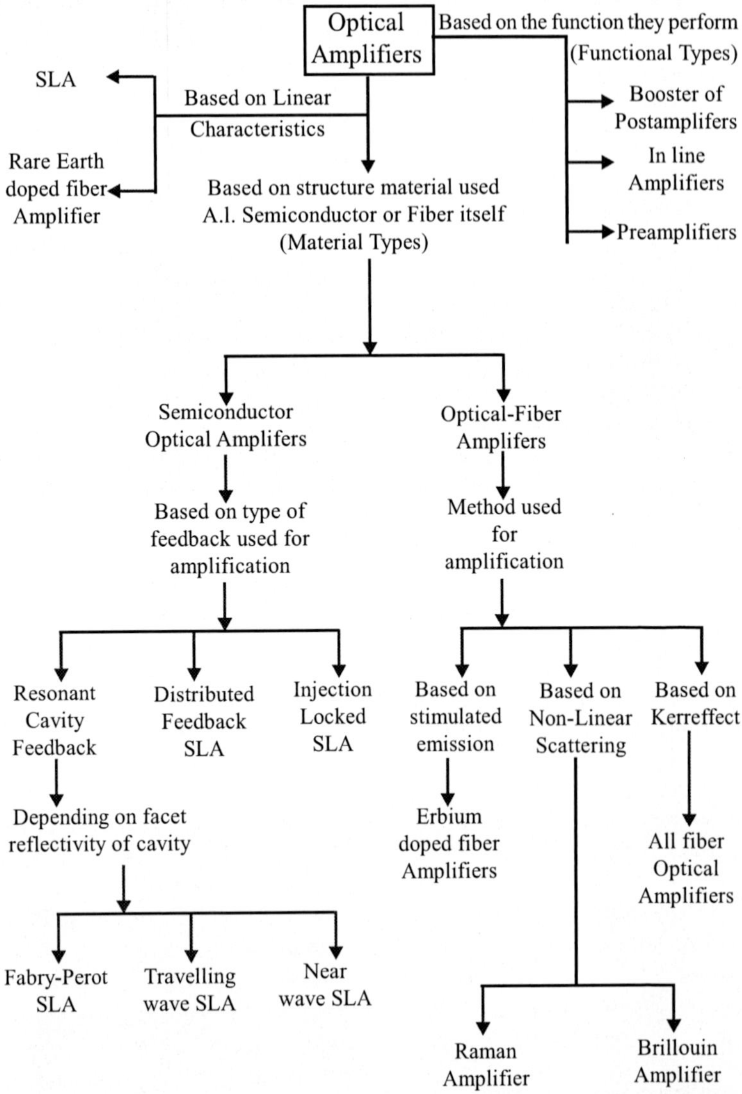

Fig 2.2 Different Types of Optical Amplifier

optical gain blocks, optical receiver preamplifiers[7], optical gates, pulse shapers, routing switches[8] and even as booster amplifiers in transmitters[9].

2.2 Optical Amplifiers

Optical amplifiers operate solely in the optical domain with no interconversion of photons to electrons. Optical amplifier deliver at its output a linearly amplified replica of the optical input signal so optical amplifiers can be placed at interval along a fiber link to provide successive linear amplification of the transmitted optical signal. Thus optical amplifiers may be considered key components to enhance flexibility of system configuration.

Over recent years, optical amplifiers have emerged as promising network elements. By using optical amplifiers, the power budget limitations of optical passive networks may be overcome and such networks could be extended. Thus optical amplifiers play a significant role in light wave system technology having speed greater than 10 Giga bit per second[10]. Additional optical amplifiers support the use of WDM, whereby signals of different wave lengths are combined and transmitted together on the same transmission fiber.

2.3Classification of Optical Amplifiers

Broadly speaking, Optical amplifiers are classified on the basis of material used, semiconductor or the fiber itself, Therefore two broad categories of optical amplifiers exist

(i) Semiconductor Optical Amplifiers

(ii) Fiber Amplifiers

Further, these amplifiers are developed on different bases whose classification tree is presented and a brief introduction of each type is discussed.

2.4 Semiconductor Optical Amplifiers

SOAs have basically the same configuration as semiconductor Laser diodes, with biases below the lasing threshold. In SOAs, a semiconductor junction is formed with one part p-type and the other n-type Material chosen is gallium-Arsenide (Ga As) so that when the junction is forward biased, the carriers are injected into each others region where the recombination takes place and the energy is released in the form of light radiation. The amplification of the radiation is found in the narrow region near forward biased p-n junction.

If the emitted radiation photons released during recombination, travels in the plane of junction then amplification is achieved. Further to increase the gain, the radiation must travel to and fro so an optical resonator called cavity containing p-n junction is formed. In essence SOA is a Laser diode without, or with very low, optical feedback.

2.5 Amplification Process

When current is injected through p-n junction of Ga-As material by means of ohmic contacts attached to the top and bottom faces, the junction emits radiation. The population inversion necessary for amplification of the radiations is found in the narrow region near forward biased p-n junction. The front and back faces are polished parallel to each other, thus forming an optical cavity. These lasers are called injection lasers, since the laser action is achieved by charge carriers injected into a semiconductor diode.

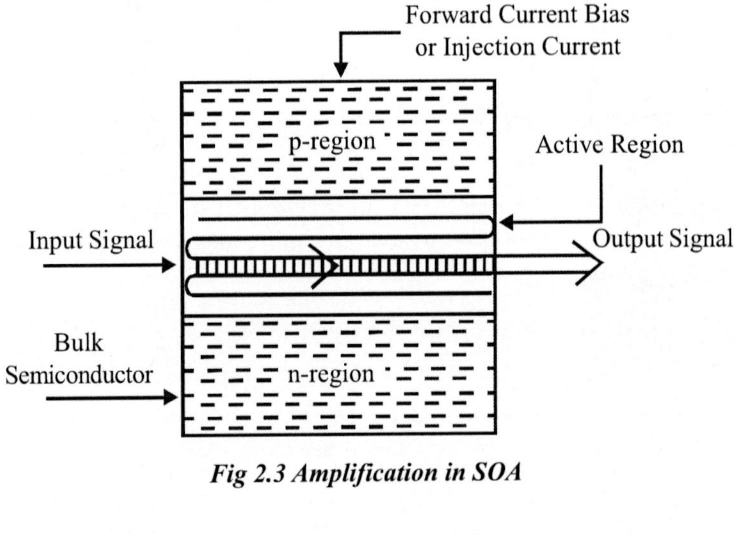

Fig 2.3 Amplification in SOA

In fact a semiconductor laser amplifier has similar structure to that of a light emitting diode (LED) and an injection laser (ILD). The same diode can behave sequentially as a LED, an amplifier and a laser depending upon the magnitude of the bias injection current. The diode has cleaved crystal facets at both ends. These facets act like a plane-parallel, partially reflective mirrors.

When injection current is relatively low, the active region of the device begins to glow with light spontaneously emitted over a broad spectrum of wavelengths. In this condition it act as an LED. Any light inserted into one end of the diode from an external source will be absorbed by the crystal[7]

As the injection current is gradually increased, there comes a point at which the diode will turn transparent to the incoming radiation. That is light inserted into it from an external source will no longer be absorbed but will pass through the diode and be emitted out the other end. This transparency point marks the threshold of amplification; the current at which the diode begins to act as an amplifier. At current levels above the amplification threshold the diode has net gain, intensifying the incident light. This gain increases as the injection current increases. In addition, the diode is still spontaneously emitting its own light and amplifying these spontaneous emission. This amplified spontaneous Emission (ASE) has some what narrower spectrum than the LED emission and is the principal source of noise of the amplifier.

As injection current is further increased the gain begins to saturate and a further increase in injection current no longer results in change in gain. The saturation point depends on the composition of the material. However if the diode has some feedback mechanism, such as partially reflective end mirrors, the gain will not saturate. Instead at a second threshold current, the net gain exceeds the transmission losses through the end facets. At this lasing threshold current, internally generated light will travel back and forth between the two crystal facet mirrors, stimulating further emission with each pass. A part of radiation will emerge out through the end mirrors. This emitted light has a fairly narrow spectrum. Thus the diode has changed from an amplifier to a laser.

2.6 Fabry Perot Semiconductor Laser Amplifiers

Depending upon the facet reflectivity, various semiconductor laser amplifier may be designed. For example when the input and output laser facet reflectivity are same and lie between 0.01 to 0.3[8] then a fabry perot semiconductor laser amplifier is obtained. FPA is operated biasing below the normal lasing threshold current, and light entering one facet appears amplified at the other facet together with inherent noise. The device is very sensitive to fluctuations in bias current, temperature and signal polarisation. However, because of the resonant nature of FPAs, combind with their high internal fields, they are used within non-linear application e.g. to provide pulse shaping and bistable elements.

2.7 Travelling Wave SLA

When the facets of cavity exhibit zero reflectivity in theory then a true travelling wave SLA is obtained. To form a TWSLA antireflection coating may be applied to the laser facets to reduce or eliminate the end reflectivities. This is achieved by depositing a thin layer of silicon oxide or silicon nitride on the end facets such that the reflectivities are reduced to 1×10^{-3} or less. Such a device become TWA operating in the single pass amplification mode in which the fabry perot resonance in suppressed by the reduction in facet reflectivity. This has the effect of substantially increasing the amplifier spectral bandwidth and it makes the transmission characteristic less dependent upon fluctuations in bias current, temperature and input signal polarisation.

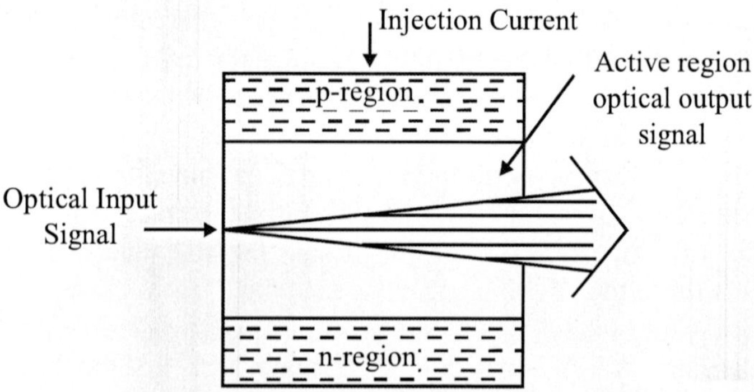

Fig 2.4 Travelling Wave SLA

Hence TWA proves superior to the FPA particularly for linear applications and also provides advantage in relation to both signal gain saturation and noise characteristics[11]. Moreover, antireflection coatings on facets have the effect of increasing the lasing current threshold and so in practice such SLAs are operated at currents far beyond the normal lasing threshold current.

2.8 Near Travelling Wave SLA

How ever in practice even with the best antireflection coatings, some residual facet reflectivity remains, for example a low reflectivity of 1×10^{-4} =0.0001 is obtained at λ=2.5mm. Thus in practical TW amplifier the zero reflectivity condition is never achieved. There exists a small value of mirror reflectivity so these devices are referred as Near TWSLA[8]

2.9 Distributed Feedback SLA

In Fabry-Perot Laser Amplifiers, the reflections at the end mirrors provide requisite feedback to obtain amplification. However there can be semiconductor Laser amplifiers in which amplification occurs without resonating mirrors. The necessary feedback in these lasers is provided by the spatial variation of the refractive index of the laser medium. These lasers are termed as Distributed Feedback (DFB) Lasers because the feedback is provided through out the entire laser medium.

The principle of distributed feedback laser is based on bragg's Law, which states that the light waves reflected from different planes in a crystal will be in phase $2d\sin\theta = m\lambda$ (Bragg's Law)

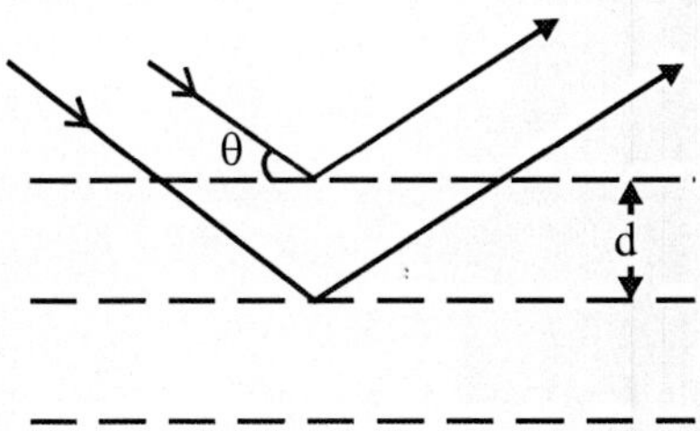

θ = glancing angle λ = wavelength of radiation
d = distance betweenn successive planes m = an integer

Fig 2.5 Bragg's Reflection

Suppose the refractive index of an active element is spatially modulated along its optical axis

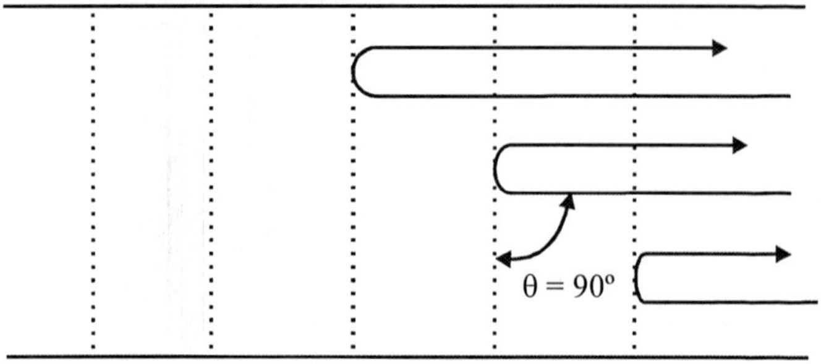

Fig 2.6 Reflection in a spatially modulated medium

In the diagram above, the dashed lines represent the reflecting planes resulting from the spatial modulation The normally incident ($\theta=90^0$) light waves will be reflected back at the planes, if the condition,

$$2d\eta= m\lambda \text{ is met.}$$

Where η= refractive index.

The energy of the wave propagated in one direction is continuously fedback in the opposite direction by Bragg scattering, thus producing a highly selective feeback.

2.10 Injection Locked Laser Amplifier

In a semiconductor DFB laser a thin film of Ga As is employed. However the spectral line widths associated with distributed feedback Laser is too broad (of the order of 5 to 50 MHz). So the Injection locked semiconductor Lasers were developed. In ILL amplifier, a sufficiently strong optical signal from an external source is injected[12] such that the frequency and phase of the output from the primary semiconductor laser will follow (lock) those of the injected signal. This is called injection locking technique.

2.11 Performance Characteristics

Amplifiers performance is measured in term of its signal gain and bandwidth. Alternatively, square root of gain and bandwidth product is used as the measure of amplifiers performance. In SOA, the gain is a function of the internal structure of the device and mirror reflectivties, the injection current and the device's physical length. The internal gain coefficient, that is, the gain per unit length within the crystal, increases linearly with the magnitude of injection current. The single pass gain of amplifiers, which is a function of device length, increase exponentially with the increase in device length. As a TW amplifiers reflects no energy from its facets, its signal gain is equal to the single pass gain in the amplifier cavity. In FP amplifiers, as the injected light is reflected back and forth between the mirrors, the effective length of the device and hence the gain of the amplifier is enhanced. The TW amplifiers have large band width in comparison to FP amplifiers.

A conventional FP laser amplifier usually has poor frequency stability and operates in many modes because of its wide gain spectrum and low cavity Q. Injection locking techniques can be used for spectral control and frequency stabilisation of the FP laser amplifier. A SLA which is locked to another frequency stablised laser is known as Injection locked laser amplifier. Injection locking helps to achieve single mode operation under high speed modulation, reduce line width, suppress partition noise and reduce frequency chirping.

The wide spectral band widths that may be achieved using high quality antireflection facet coatings on TWAs are in the region of 50 to 70 nm. However in comparison to FPAs, such device require significantly higher bias currents. The narrow spectral band width of FPAs provides in herent noise filtering which is not obtained with TWA and therefore they are subject to increased levels of noise.

The residual facet reflectivity in TWAs introduces a problem of backward gain when used as amplifiers within optical fiber communication system[13]. In systems with cascaded amplifiers, optical isolators may therefore be required to avoid the interaction of backward signals between the devices.

Another, very important characteristic of SLAs is noise since it largely determines the maximum number of device which can be cascaded as linear repeaters within an optical fiber communication system[11]. The overall noise generated by an SLA comprises signal spontaneously beat noise, spontaneous-spon taneous beat noise, spontaneous emission shot noise and amplified shot noise[14]. The beat noise component occur between the signal and the spontaneously emitted photons as well as between spontaneously emitted photons themselves. In addition the former noise component tend to dominate in the higher power operating region required for repeater applications. It is clear that, unlike an electronic amplifier, the noise output from an SLA is a function of the signal intensity.

The gain of FPA is sensitive to changes in temperature and signal polarisation. So is the case with TWAs. For example, at an operating wavelength of 2.5mm the gain decreases by around 3dB when the temperature of a TWA of length 500mm is increased by 5^0C[8]. Moreover, although a decrease in temperature increases the device gain, it also increases the passband ripple when there is residual reflectivity.

2.12 Fiber Amplifiers

It is found that Semiconductor Laser Amplifiers are prone to variations resulting from temperature changes, ageing and pump power. Therefore development of fiber amplifiers took place where in these effects are less significant.

In fiber amplifiers the optical gain is provided by either stimulated Raman Scattering, Stimulated Brillouin Scattering or by rare earth doped materials. Accordingly fiber amplifiers are classified as

(i) Raman Fiber Amplifiers

(ii) Brillouin Fiber Amplifiers

(iii) Rare Earth Doped Fiber Amplifiers

Using Raman and Brillouin effects, it is possible to build distributed, not lump, amplification of an optical signal. In this the same piece of fiber serves as a transmission and amplification medium simulataneously. In all Fiber amplifier devices the spectral band widths and center wavelengths are largely defined by the atomic structure and mechanical geometry.

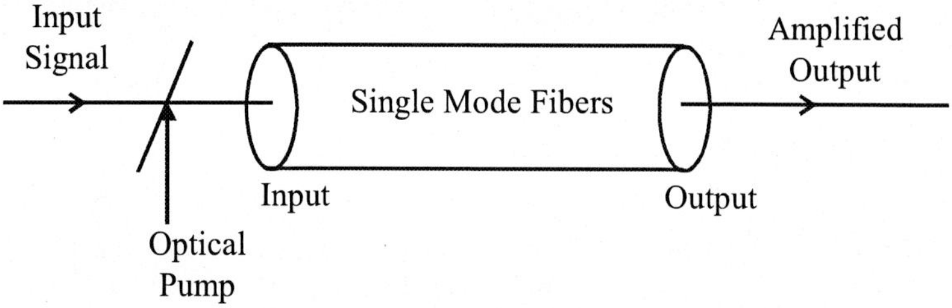

Fig. 2.7 Operation of Fiber Amplifier

A general representation of fiber amplifier operation is shown in figure. It consists of an optical pump which is a high power semiconductor laser, and an optical coupler where in the incoming optical signal and the pump signal is combined and then launched in the single mode fiber. Single mode fiber is the gain medium where in the gain is provided. The amplified optical signal is emitted from the other end of the gain medium

Materials for Optical Fiber

- Core Materials : SiO_2 , GeO_2
- Cladding Materials : SiO_2 , P_2O_5 , F, B_2O_3
- Rare Earth Materials : Er, Nd, Sm, Tb

2.13 Amplification Process in Fiber Amplifier

In the fiber amplifiers the external energy to the system is delivered optically and not electrically as in SOAs. Pumping is done with a laser diode raaiating powerful light at a wavelength other than an input information signal's wavelength. Both the optical information and optical pumping beams are put in the same fiber through a coupler. These two beams propagate together along the section of the fiber, where the information signal is amplified while the pumping signal loses its power. In a sense, then pumping light gives its power to an information signal and "dies". A second coupler removes residual pumping light from the transmission fiber.

2.14 Raman Fiber Amplifiers

RFA are based on Non-Linear Stimulated Raman scattering in which the incident photons produce phonons, A phonon is an elementary particle analogous to a photon but differs from a photon in its quantum properties.

In the process of producing phonons which is a quantum of an elastic wave in a crystal lattice of high optical frequency, also has a scattered photon. The incident wave canbe seen as a kind of pump wave when signal from pump laser is injected into the fiber, then SRS causes amplification of the signal.

Thus Raman amplification is a process by which Signal light is amplified Via non linear interaction with pump light and optical phonons. The frequency relationship of $n_{Pump} = n_{Sig} + n_{Phon}$ is satisfied in the process.

Two proceses are posible in the Raman interaction involving the pump, signal and phonon.

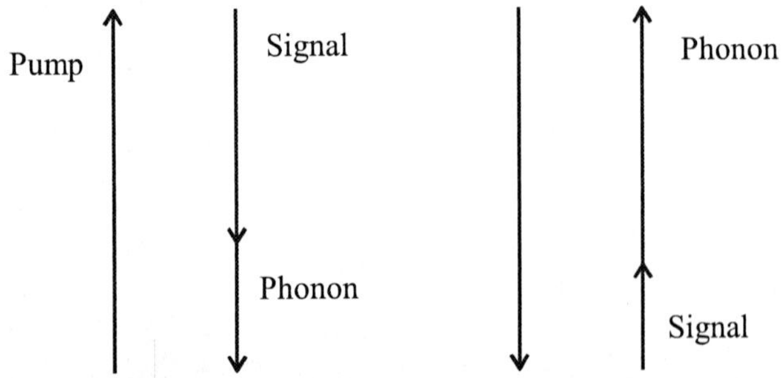

(a) Signal Photon (b) Signal absorption

(a) A pump photon is absorbed and signal photon and an optical phonon are created due to stimulated emission.

(b) A signal photon and an optical phonon are absorbed and a pump photon in created due to stimulated absorption.

A schematic diagram of Raman Amplifier in shown in fig. In coming optical signal is assumed to be launched at z=0 and travel in +z direction in the fiber. Pump light is injected at z=0 for forward pumping and at Z=L for back backward pumping, where L is the length of the optical fiber. In fig(b) the relation between pump and signal frequencies is shown. Dotted line shows a Raman gain profile that is assumed to be rectangular, fg is the raman gain band width. The signal frequency can be chosen freely anywhere inside the Raman gain band width Δfg. The raman gain depends upon the pump power and fiber length.

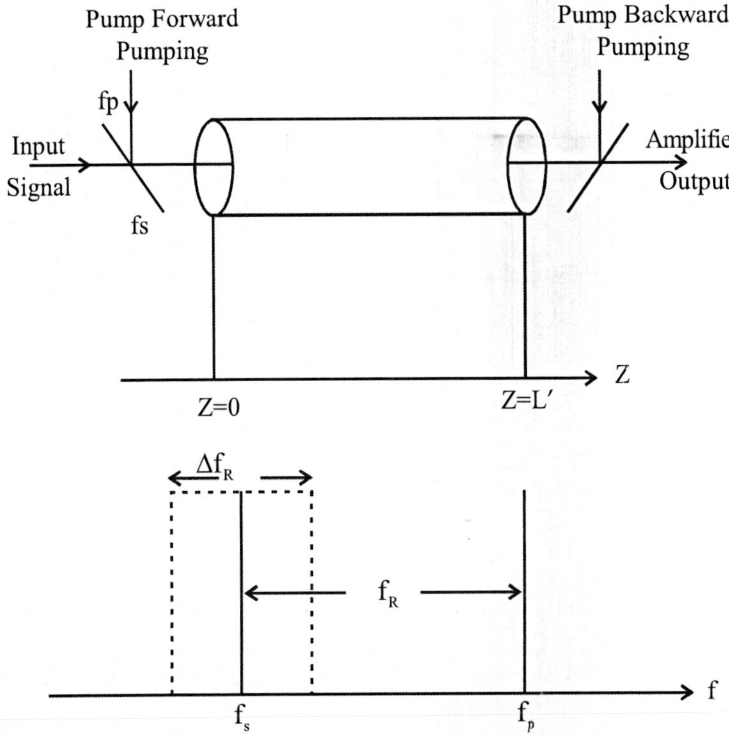

Fig. 2.8 Principle of Operation of Raman Amplifier

2.15 Brillouin Fiber Amplifiers

BFA are based on Non-Linear Stimulated Brillouin Scattering (NLSBS) in which the incident photons produce phonons of acoustic frequency. Thus although in SRS the scattered light moves mostly forward and the phonons associated with this process are optical ones; SBS scattered light moves backward and the phonons associated with it are acoustic.[15]

2.16 Comparison of Different Optical Amplifiers

	Fabry-Perot Semicounctor Laser Amplifier	Travelling Wave Semiconductor Laser Amplifier	Injection Locked Semiconductor Laser Amplifier	Fabry-Perot Raman Laser Amplifier
Small Signal Gain	25-35 dB	~ 20 dB	20-30 dB	20-30 dB
Saturation Output Level	–10 to 0 dB	0-10 dB	10-20 dB	~ 20 dB
Bandwidth	1-3 GHz	~1 THz	0.5-1 GHz	~ 1 THz
Noise Figure	3 dB	3 dB	3 dB	3 dB
Excess Noise Bandwidth	Medium	Large	Small	Medium
Input/Output Coupling Loss	Medium	Medium	Medium	Low
Gain Stability against Frequency Deviation	Medium	High	Low	Medium
Merits	Less Spontaneous Emission Noise	Broad Bandwidth	Less Spontaneous Emission Noise	Broad Bandwidth
Demerits	Small Output Power	Need for Narrowband Optical Filter	Need for Frequency Stabilised Source	Need for Reliable Pump Laser in 1500 nm range

2.17 Rare Earth Doped Fiber Amplifiers

Rare Earth material e.g. Erbium is doped in fibers to obtain amplification. In Erbium Doped Fiber Amplifiers EDFAs, the piece of fiber working as an active gain medium is heavily doped with ions of Erbium (Er). Amplification in EDFA occurs through the mechanism of stimulated emission. Energy from the pumping signal excites erbium ions at the upper energy band. The information signal stimulates transition of the excited ions to the lower Energy band. These transitions result in the radiation of photons with the same wavelength, the input signal has.

2.18 All-Optical Fiber Amplifiers

All-Optical Fiber Amplifiers has band width in THz Range [16]. This amplifier in principle is an integrated Mach-Zehnder Interferometer where both arms are pumped with light generated from the incoming signal as shown in the sketch below

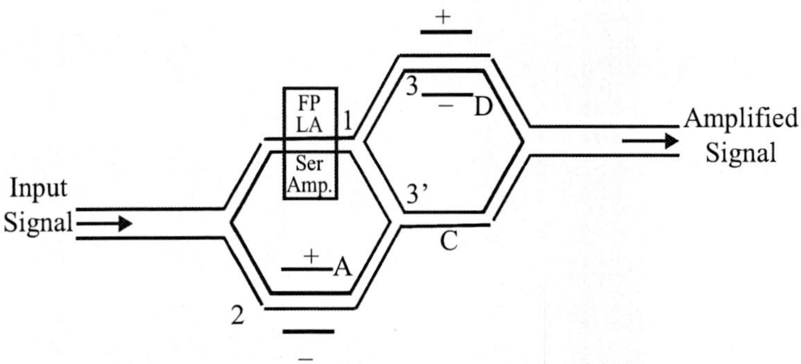

Fig.2.9 Basic Configuration of All Fiber Amplifiers

In one of the branches is a FP laser amplifier. The laser amplifier output is large and of constant amplitude, however the phase of the incoming signal is conserved, if these fluctuations are slow. This has the advantage that the phase relation between point 1 and 2 becomes fixed in time, Hence the phase at 3' and 4 can be made equal with the help of electrodes in arm A which changes the index of refraction by means of dc Kerr-effect. Signal in arm C is the sum of signals at points 3' and 4 can be made equal with the help of electrodes in arm A which changes the index of refraction by means of d.c. Kerr-effect. Signal in arm C is the sum of signals at points 3' and 4. Optical length of arm 'c' varies is proportion to the amplitude of the original signal, while the other arm D is unaffected by the rapid amplitude Variations.

So an additional pair of electrodes in arm D is required to compensate for imperfections in the arm C and D and phase distortion at the Y-junction So that the optical path of C and D are equal when signal field is zero. The two are then united at 5 with symmetric Y-junction. It has been reported that the resultant amplitude is proportional to the incoming signal amplitude with a distortion less than one percent.

2.19 Phase Sensitive and Phase Insensitive Amplifiers

The difference between a phase sensitive amplifier and a phase insensitive amplifier is analogous to that between an optical homodyne detector and a heterodyne detector. An optical homodyne detector performs on input signal without any additional noise but detects only one quadratur component, while an optical heterodyne detector measures two quadrature components simultaneously but with additional noise.

Laser amplifier and a non-degenerate parametric amplifier are examples of a-phase insensitive amplifier, while a degenerate parametric amplifier realizes noise-free phase sensitive amplifiers.

2.20 Conclusion

All-optical amplifiers used in long haul communication have been discussed here. Wide Range of amplifiers provide the professionals an opportunity to select the one needed for its application. But the current evolution telecommunication system to higher information capacity Involving Local Area Network needs to be attended. The Challenges of optical amplification in Area Network are deployment cost, Capacity growth profile and operational expenditure. Architectural decision of long haul optical amplifiers are not the best solutions so the specific application based amplifiers are required to be designed.

Further, the data speed of optics is yet to be achieved due to electro-optic signal processors still limiting the capacity of photons by in between converting them to electrons. This is further aggravated in terms of power budget limitations because there is a power limit to launching the light in optical fiber. Therefore need of amplification at processors level arise to compensate for power budgetary deficit or loss.

References

[1] John M Senior, "Intergated Optics", Optical Fiber Communications, Second Edittion, PHI, 1996, pp 531-532

[2] Yoshihisa Yamamoto & Kyo Inone; "Noise in Amplifiers", JLI, Vol 21, No. II, November 2003, pp 2895-2915

[3] Atkins CG; Cotter D; Smith DW; Wyatt R; "Application of Brillouin amplification in coherent opitcal transmission", Electron Lett; 22(1986) pp 556-558

[4] B Jarklev A; "Optical Fiber amplifier, Design and system applications", Artech House Publishers' 1993, pp 218-230

[5] Silva Da VL' "Remotely pumped Erbium doped fiber amplifiers for repeater less submarine systems"; IEEE Photonics Techn Letter 7(1995)9, pp 1081-1083

[6] Taga, H; Recent progress in amplified undersea system; IEEE JLT, 13(1995)5, pp 829-840

[7] Dr. Narsingh; "Performance Evaluation of Receivers in digital coherent OFC system" Ph.D Thesis, pp 55-66, 1989, IIT, Delhi.

[8] MJ O' Mohony Semiconductor laser optical Amplifiers for use in future fiber systems", JLT; 6(4), 1988, pp 531-544

[9] Yonggyoo Kim and Jichai Jeong, "Transmission using Semiconductor Optical Amplifierss as booster amplifiers", JLT, Vol. 21, No. 2, February 2003, pp 476-495

[10] DC Agarwal, Fiber Optic Communication", Wheeler and Sons Pub.; Optical Amplifierss, pp 233-241

[11] T. Saitoh and T Mukai, Recent progress in Semiconductor Laser amplifiers", JLT' 6(11), 1988, pp 1656-1664

[12] R. Lang; "Injection Locking properties of a Semiconductor Laser; IEEE J of Quantum Electronics, QE-18(6), pp 976-983, 1982

[13] I.D. Henning; MJ Adams and JV Collins; "Performance predictions from a new optical amplifier model", IEEE J. of Quantum Electronics QE-21, 1985, pp 609, 613

[14] John M Senior, "Optical Amplifiers", Optical Fiber Communications, PHI, 2003, pp 514

[15] D Jafar, K Mynbaev and Lowell L Scheiner; "Repeaters and Amplifiers", Fiber Optic Communication technology; Pub-Pearson education Asia, 2002, pp 508-509, 523

[16] Manneberg G, Kull M; and Lind Quist SE; "An all optical fiber amplifier for THz Signals, A novel proposal' 1987, JLT, Vol. LT-5, pp 236-242

3

Noise and Disturbances in Optical Amplification

3.1 Introduction

Noise is an unwanted energy which creeps into amplified signal in the process of amplification. Light radiated by a laser diode fluctuates in its intensity and phase even when biasing current is ideally constant[1]. These disturbances require adequate precaution in designing an optical amplifier. Broadly these can be classified as optical and electrical noise.

Optical noise is the noise measured in an optical field by optical measurement devices whereas, electrical noise is a characteristic of an optical amplifier measured at a point just beyond the amplification after converting the optical signal into electrical signal. Coherent optical communication system are in particular very much sensitive to the noise of transmitter and local laser [2] Laser output radiation fluctuations are caused mostly by spontaneous emission which are random in nature and are called laser noise. These are termed as Laser Intensity noise, Laser phase noise, Laser frequency noise and so called phase-change noise.

3.2 Mathematical Representation

In an ideal Laser the spontaneous emission does not occur. The power spectral density or emission spectrum is given by-

$G_a(f) \longrightarrow G_o(f_i)$ (ideal emission spectrum '0' noise)

$f_a \longrightarrow f_o(f)$ (Single frequency)

$f_a \longrightarrow f_0$ (Constant phase)

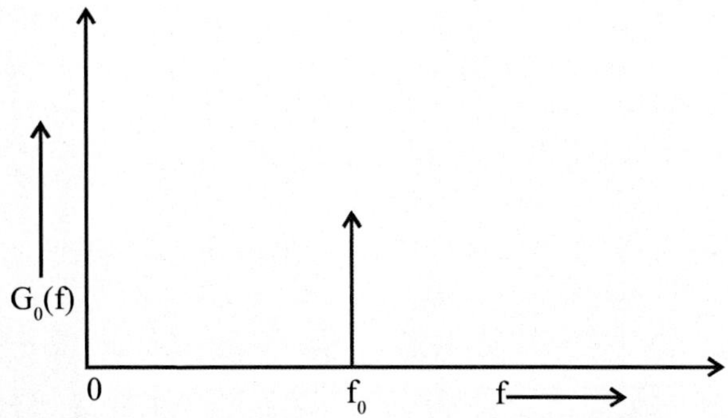

Fig 3.1 Ideal Emission spectrum is Dirac-Delta Function

But the time dependene of the electric field at the output of ideal single mode laser can be described by the equation $E_0(t) = \hat{E}_0\pi^{i2\times f_0 t} = \hat{E}_0\pi^{j\phi 0}\pi^{j2\pi f_0 t}$ Which represents a harmonic oscillation[3].

But the spontaneous emission yield a time-varying amplitude $|E(t)| = \hat{E}(t)$ and, in addition, a time varying phase $\phi(t)$. As a result, laser-emission spectrum substantially broaden.

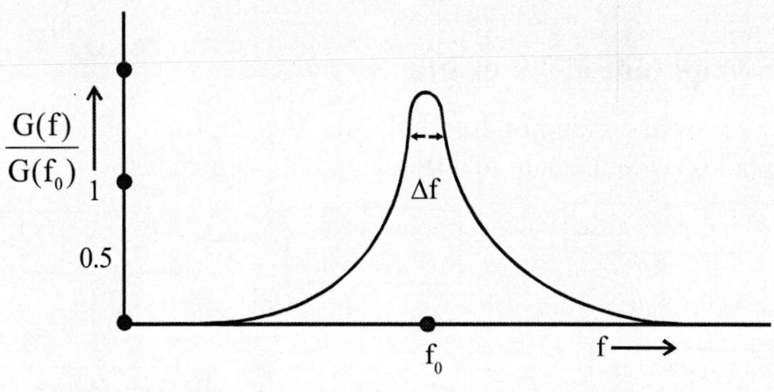

Fig. 3.2 Emission Spectrum of Real Single Mode Laser

The electric field of the transmitted laser wave, which is disturbed by laser noise, can be expressed as

$$E(t) = \hat{E}_{(t)} e^{j[\phi+\phi(t)]} \cdot e^{j2\pi f0(t)}$$

The time varying random phase f(t) is called the laser phase Noise. And the random amplitude |E(t)| as laser amplitude or Laser Intensity Noise. Laser excess noise normally described by relative intensity noise describes the noise generated in addition to shot noise[4].

A quantitative measure of Laser Intensity Noise is

$$RIN = \frac{\Delta\overline{P}^2}{\Delta\overline{P}_0^2}$$

where $\Delta\overline{P}^2$ = mean Square Intensity fluctuation in spectral density of the light output.

$\overline{P}_0$ time average of Laser light intensity.

The fundamental measure to assess the strength of laser noise is given by the spectral laser line width $\Delta fa \rightarrow \Delta f$ which is termed as 3-db line width, Spectral Half width or line width at FWHM given by

$$G\left(f_0 \pm \frac{\Delta f}{2}\right) = 0.5G(f_0)$$

3.3 Amplitude or Laser Intensity Noise

In semiconductor Laser diodes the addition of noise is explained with the help of a Block diagram drawn below-

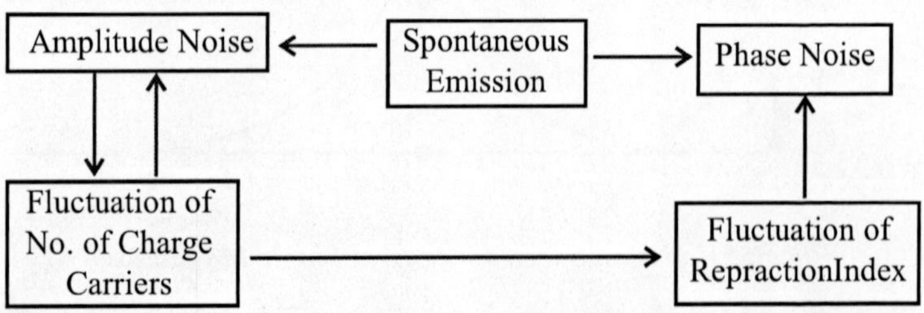

3.4 Noise in Optical Amplifiers

When a weak signal is amplified the noise associated with the signal is also amplified. If an amplifier does not add any excess noise to the amplified signal then the SNR is preserved. In many cases, however, an amplifier has independent internal noise that i is added to the output signal. Optical Amplifier magnifies the signal noise along with the signal itself, but actually it does much more than that. An optical amplifier generates its own noise. Therefore, an amplifier changes the SNR of both the input and output signals. The noise performance of an optical amplifier is quantified through "Noise Figure (F_n)"

$$F_n = \frac{(SNR)_{IN}}{(SNR)_{OUT}}$$

F_n is often referred to as figure of merit.

When amplifier adds noise to the output signal then SNR is said to be degraded. Noisiness of a linear amplifier is expressed by Noise figure.

The fundamental quantum limit of the noise figure of a linear amplifier is F = 2 (3dB) which can be reduced to F = 1 (0dB). Just like a homodyne detector has a 3 dB better sensitivity than a heterodyne detector. This specific noise free amplifier having F=1 is called a phase sensitive amplifier because in this only one quadrature (cos wt or sin wt) is amplified instead of two[5].

3.5 ASE-Amplified Spontaneous Emission

The noise generated by an active medium of an optical amplifier is caused primarily by ASE.

The physics of this phenomena is that the vast majority of excited carriers are forced by stimulated emission to fall to a lower level, although some of these carriers do so spontaneously. When they decay, they radiate photons spontaneously. The spontaneously emitted photons are in the same frequency range as the information signal, but they are random in phases and direction. The spontaneously emitted photons that follow in the direction of the information signal are amplified by an active medium. These spontaneously emitted and amplified photons constitute ASE. But they are random in phase, so they do not contribute to the information signal but generate noise within the signal's band width. ASE noise is spread over a wide wave length range.

The average total power of amplified spontaneous emission(P_{ASE}) is given by the equation.

$$P_{ASE} = 2\eta_{sp}\, hf\, G\, BW \text{ [6]}$$

Where η_{SP} = spontaneous emission factor or population inversion factor

hf = Photon Energy

G = Amplifier gain

BW = Optical bandwidth of the amplifier.

This formula clearly expresses the idea that the greater the spontaneouse emission, quantified by η_{Sp}, the greater too will be the amplified spontaneou emission(ASE).

3.6 Double Rayleigh Scattering Noise

It occurs in fiber Raan amplifiers. Rayleigh scattering is an unavoidable property of the transmission fiber which is contributed by two phenomena.

(i) Due to Amplified Spontaneous Scattering Counter propagating with respect to the signal.

(ii) Due to homodyne cross talk the above phenomena is aggravated so it's called Double Rayleigh scattering (DRS).

Double Rayleigh Scattering originates an incoherent multiple path interference (MPI). Which is a noise generated by the beating of multiple differently delayed replicas of the signal itself [7]. This DRS can not be removed by optical filtering. So while optimising the system capacity by WDM, the signal to noise ratio must be improved by taking DRS into consideration.

Let

L = total length of fiber Raman Amplifier from 0 to Z=L

ξ and Z = the points between which DRS takes place.

K_{RS} = Rayleigh back scattering coefficient in m^{-1}

– (minus) = Sign to indicate counter-propagation

+ (plus) = Sign to indicate co-propagation

P_{SO} = Signal intput power in watts

G^{-} (o, z) = amplification gain from o to z.

K_{RS} G^{-} (o, z) P_{SO} dz = Very small back scattered power, this backscattered in z power is amplified by multiplying by = $G^{+}(Z, \xi)$

In ξ point, this contribution is back scattered once again and from ξ to z it gain the amplification = $G^{-}(\xi, Z)$

The whole DRS power in z is found by integrating ξ between zero and z. This contribution is then amplified by G^{-} (z, L), the total amount of DRS power at the fiber end, P_{DRS} sig, is thus obtained by intergrating along Z.

$$P_{DRS}sig = K_{RS}^{2}P_{SO}\int_{0}^{L}\int_{0}^{Z}G^{-}(o,z).G^{+}(z,\xi).G^{-}(\xi,L)d\xi dz \text{ [8]}$$

The scematic of the small signal model for DRS power calculation canbe drawn as [8].

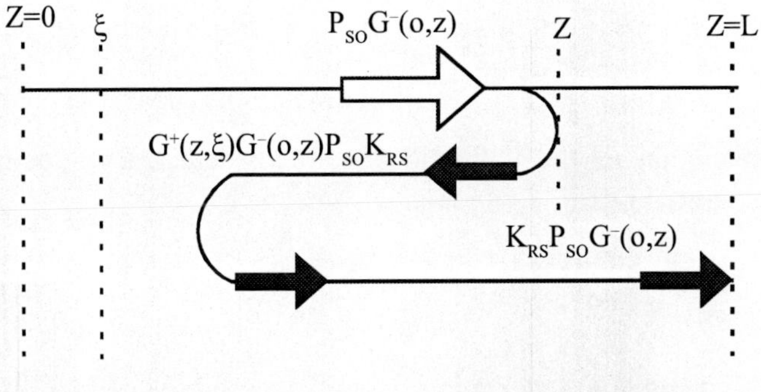

$$K_{RS}^{2}P_{SO}G^{-}(o,z).G^{+}(x,\xi)K_{RS}^{2}P_{SO}G^{-}(o,z).G^{+}(z,\xi)G^{-}(\xi,L)$$

3.7 Beat Noise

The quantum fluctuation generates spontaneous emission, and then the spontaneous emission causes signal fluctuation.

After Amplification let us consider that signal light is incident into a detector together with ASE and express the incident light as

$$E = A_s \pi^{[J\omega_{st} + \theta_s]} + \sum_K A_{sp}^{(K)} \pi^{\left[j\left(\omega_{sp}^{(K)} t + \theta_{sp}^{(K)}\right)\right]} \quad [9].$$

The First term denotes the signal where, As, ws, and qs are the amplitude, angular frequency and phase respectively. The second term denotes the ASE light. Beat Means pulsate and in optical amplifiers waxing and waning of a signal is called Beat. In the above case the mixing between the amplified signal and the ASE components, result in Beat thereby generating signal spontaneous Beat Noise. Similarly spontaneous-spontaneous beat noise is generated by mixing within the ASE components themselves.

In addition to these two noise components, there will be shot noise components produced by the signal and the spontaneous emission. Because the amplifier output signall plus. ASE when detected by a photodiode consists of desired signal component along with the beat and shot noise cmponents.[10]

In order to model the ASE that is continuously broadened in the wavelength, the optical frequency band is divided by a frequency df, and ASE in the Kth band is expressed as $A_{sp}^{(K)} \exp\left\{j\left(\omega_{sp}^{(K)} t + \theta_{sp}^{(K)}\right)\right\}$ and the summation is taken over the ASE band width when the light field E is directly detected, photocurrent, is generated as $I \alpha |E|^2$ which is written as

$$I = \frac{\eta c}{h\nu}\left\{A_s^2 + \sum_k A_{sp}^{(k)^2} + 2AS\sum_k A_{sp}^{(k)}\right.$$

$$\cos\left[\left(\omega_{sp}^{(k)} - \omega_s\right)t + \theta_{sp}^{(K)} - \theta_s\right] + 2\sum_{k<k'} A_{sp}^{(k)} A_{sp}^{(k')}$$

$$\left.\cos\left[\left(\omega_{sp}^{(k)} - \omega_{sp}^{(k')}\right)t + \theta_{sp}^{(K)} - \theta_{sp}^{(k')}\right]\right\} \quad [11]$$

when h is the detector efficiency, e is the electron charge, h is planek's constant n is the optical frequency. First term is signal power, the second term is spontaneous emission power, third term is the interference between the signal and ASE lights, fourth term is interference between ASE lights.

The phase of ASE light randomly fluctuates thus the third and fourth term fluctuate accordingly, which result in noise. Because the third and fourth terms have no correlation, so they can be individually treated. The noise due to third terms is signal spontaneous. Beat noise and the noise due to fourth term is spontaneous–spontaneous beat noise.

3.8 Interference Noise

For example in Raman amplifiers, a long length of fiber is used ass gain medium. In such a long fiber Rayleing scattering causes distributive reflection along the fiber. The reflected light is scattered again towards the original direction. Due to this double Rayleigh scattering, ASE power is enhanced and the noise figure is degraded. The double Rayleigh scattering causes. Interference noise. The signal light is scattered again in the forward direction. The original signal and the double scattered light interfere with each other at the detection stage. Since the optical length of fiber and the phase of the laser source fluctuate in time, the phase difference of the interfering lights fluctuate and the received signal fluctuates giving rise to interference noise[12].

3.9 Cross Saturation

In a multichannel transmission system when the light intensity of one channel is high then amplifier is gain saturated. In this situation signal output of one channel varies according to signal gain fluctuation induced by the modulation of other channels even when this signal input is constant. This results in cross talk induced by gain saturation in the amplifier[13].

3.10 Polarisation Dependent Gain

The amplification of TE and TM modes are different. Because of the rectangular shape and the crystal structure of the active region. The difference in gain between two orthogonal polarisations can reach 5 to 7 dB.

The problem with polarisation dependent gain stems from the fact that almost all of the deployed fiber is non-polarisation maintaining fiber.

There are several means to reduce it.

1. One way is to make the active region as square as possible in cross section.
2. Another way is to connect two SOAs in Series or in parallel to compensate for the orthogonal polarisations unequal gain.
3. A double pass through the same active region also compensates for unequal gain because a "faraday rotator" turns the polarisation of the back reflected beam at a 90^0 angle w.r.t. the beam.

The result of all these, is the Polarisation Dependent gain is reduced to 0.5 dB.

3.11 Conclusion

In this Chapter I have reviewed all possible noise in linear optical amplifiers. Noise figure was discussed to describe the performance of amplifier. The communication system employing optical amplifiers must have power budgetary compensation. Knowledge of these noise will help designers and users to select the suitable material, technology and amplifier.

References

[1] DK Mynbaev, Lowell L Scheiner; "Fiber optic communications technology;" Pearson Education, Asia 2002, pp. 399-400.

[2] W Imajuku. A. Takada and Y. Yamabayashi; "Inline coherent optical amplifer with noise figure lower than 3 d_B quantum limit"; Electron Letters, Vol. 36, Jan 2000, pp. 63-64

[3] JH Franz, VK Jain; "Reasons and Formation of Laser Noise"; Optical Communication' Narosa Publishing House, 2004, pp 45-50,

[4] Y. Yamamoto; "AM and FM Quantum Noise in Semiconductor Lasers Part-I : Theoretical Analysis"; IEEE J. Quantum Electronics, Vol. QE-19, Jan 1983, pp. 34-36

[5] AE Kelly, I.f. Lealman et. all; "Low noise figure and high gain SOA with single layer AR coating;" Electron Letters, Vol. 33, March 1997, pp. 536-538

[6] Djafar; "Semi Conductor optical amplifiers"; Fiber optical communication technology; Pearson education, Asia 2004 pp. 539-540

[7] P.B. Hanseh, L Skildsen et all; "Rayleigh Scattering limitatins in distributed Raman Pre-amplifiers"; IEEE Photonics Technology Letter, Vol.-10, No. 1, 1998, pp 159-161

[8] Paola Parolari, Lucia Marazzi et all; "Double Raylleigh sscattering Noise in Lumped and Distributed Raman Amplifiers"; JLT, Vol.12, No. 11, October 2003, pp. 2225-2228.

[9] HA Haus; "The noise figure of opitcal amplifers"; IEEE Photon Technology Lettes, Vol. 10, pp (1602-1604), Nov. 1998.

[10] VK Jain, JH Franz; "Noise in optical Amplifiers"; Optical Communation, Components and System; Narosa Publishing House, 2006, pp. 262-263

[11] Yoshihisa Yamamotoet all; "Noise in amplifiers" JLT, Vol-21, No.11, November-2003, pp 2895-2915

[12] C.R.S. Fludger and R.J. Mears, "Electrical Measurement of Multipath interference in distriubuted Raman Amplifiers", JLT, Vol-19, April-2001, pp 536-545

[13] K/ Inoue and T. Mukai", Experimental study on noise characteristics of a gain-saturated fiber optical parametric amplifier", JLT, Vol-20, June-2002, pp 969-974

4

Design of Optical Amplifiers for Metro and Access Application

4.1 Introduction

Advent of optical amplifiers found its initial application in booster configuration to extend the range of 1550 nm transmission links [1]. Semiconductor laser pumped erbium doped fiber amplifiers consisting of a short length of optical fiber containing a low concentration of erbium ions, and a semiconductor laser that excites these ions to provide gain in the 1550 nm wavelength region can boost the power of light wave signals without the need for optoelectronic conversion and subsequent electronic application as in conventional light wave repeaters. This simple and elegant function, which is independent of the operating data rate and can simultaneously accommodate many WDM channels was being employed with booster amplifiers to double the capacity of installed fiber routs under coarse wavelength division multiplexing architecture offering simple and cost effective means to access the Vast transmission capacity inherent in single mode optical fibers.[2]

In a WDM system many (4,8,16 or 32) wavelength carrying data are optically multiplexed and simultaneously transmitted through a single fiber. The received signal with many wavelengths is optically de multiplexed into separate channels, which are then processed electronically in a conventional form. Such a WDM system needs transmitters with many lasers at specific wavelengths

Cascades of EDFA were under investigation for undersea use; first in single wavelength configurations then soon it became apparent that multiple wavelengths could be supported with cascaded OA systems leading to the application of Dense Wavelength Division Multiplexing and EDFA technology [3].

4.2 Metro Access Network

The present work explores the suitability of optical amplifiers in metropolitan area network. MAN is designed to serve a certain geographical region of metropolitan city. Typically a metropolitan area network spans distance upto a few tens of kilometers to a hundred kilometers depending on the size of the city. In general the metropolitan transport system comprise both the regional (core) as well as access (edge) networks. These systems provide a bridge between the wide Area network (WAN) and Local Area network (LANs) supporting a variety of services to different users. The metro transport system differs from a long haul system in many ways such as span length, network architecture, scalability, switching, type of service and cost.

4.3 DWDM for Access Application

Once the success was achieved in optically amplified DWDM systems in long haul applications, it was realized that terrestrial networks and undersea cable systems that employ optical amplifiers can have significantly increased transmission capacity, networking functionality and operational flexibility ,all potentially at lower cost than with opto-electronic regenerators. These advantages can be exploited especially to provide greater network reliability, which is highly needed by service providers. But the introduction of DWDM technology into metro system incurs extra cost due to the use of expensive fixed ITU grid DFB Lasers. The use of tunable lasers as spares in DWDM metro system however does reduces the operating expenses [4].

So, initial DWDM deployments were used in backbone networks where networks reconfigurations were few. The economics of DWDM were next applied to Metropolitan Area Networks where configurations were much more frequent.

Wavelength routing and optical protection switching have become commonplace in these networks [5]. This forces additional complexity onto the OA designers to suppress transient cross talk between wavelengths. As optical amplifiers are pushed further toward the edges of the network, cost and complexity become key concern. In the early days it was all about deploying the maximum bandwidth from point A to B with the lowest overall capital cost. Now carriers are much concerned with first development cost, capacity growth profile and operational expenses, to meet all these requirements DWDM finds its place infront in Access networks..

4.4 Metro Transport Infrastructure

In recent years optical transport system have evolved from a single channel SONET/SDH transmission to multiwavelength optical networks [6]. However, the optical networks and single channel SONET/SDH systems differ in terms of capacity expansion and channel routing. In SONET/SDH systems, the enhancement of the capacity is achieved by adding a new transmission system in parallel over different fibers. On the other hand, the capacity enhancement in optical networks is accomplished by simply adding new wavelengths over the same fiber. In addition, channel routing is performed by using wavelength routing algorithms in optical networks as opposed to the use of time slots in single channel SONET/SDH systems.

The first step toward optical networking is the utilization of a wavelength division multiplexing scheme. The use of WDM stems from the internet data traffic and other high bandwidth services. The high bandwidth services include high resolution graphics, video-on-demand and multimedia communications. Research projects such as multiwavelength optical networking (MONET) and wideband all-optical-networks (AON) laid a firm foundation for the development of next generation optical networks. Other projects such as optical regional access network (ORAN) and Rainbow [7] have addressed the issues involved with the design of metro regional as well as access networks. The metro transport system architectures are broadly classified into two types : (1) ring topology (2) mesh topology. First let us look at the ring topology. As discussed earlier, the first generation metro transport systems were based on a traditional SONET/SDH ring architecture, primarily designed for voice transport based on time division (TDM) technology. Figure depicts a typical ring architecture of a metro transport system [8].

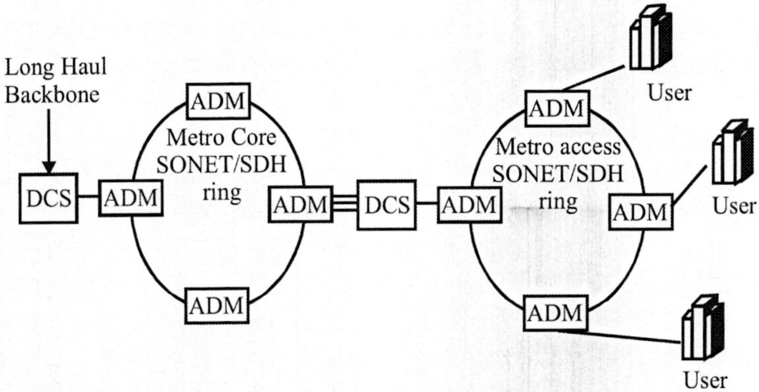

Fig. 4.1 Traditional SONET/SDH ring based metro transport system architecture.

As can be seen from Figure the system architecture consists of both metro core and access rings. The metro core ring is shared by all users, whereas the metro access ring interfaces with a variety of users such as local exchange carriers, businesses, service providers and university campuses. In a traditional SONET/SDH ring architecture, a typical node Consists of SONET add/drop multiplexers (ADMs). The ADMs allow traffic to enter and exit the ring. The digital cross connects (DCS) are used to switch the individual data streams from the long-haul backbone network or to interconnect two or more SONET rings. A DCS device utilizes an electric switch fabric and O-E-O converters [9].

In recent years, as the data traffic increased, the traditional SONET/SDH ring infrastructure became ineffective due to several layers of add/drops and DCS elements. Hence, the introduction of DWDM technology into metro core ring which resulted in the next generation metro optical networks. In addition to capacity enhancement, bit rate and protocol transparency switching efficient management control of the network operations were achieved. In DWDM based ring architecture, a node is comprises of optical add/drop multiplexer (OADMs) and optical switches[10].

An OADM can be either fixed or reconfigurati le. In a fixed OADM, a fixed wavelength will be either added or dropped into or from the data stream. On the other hand, a reconfigurable OADM deals with many wavelengths, which have to be either added or dropped at any node on the ring. In addition to the use of OADMs, optical cross-connects (OXCs) are being utilized in DWDM metro networks at the central offices to interconnect the rings, as more wavelengths and fibers are added to the nodes. The OXCs provide a variety of network functionalities such as wavelength routing, bandwidth management and network restoration. The OADMs and OXCs are the basic building blocks of a DWDM optical transport network. The key features of some of the practical DWDM metro transport systems are summarized in Table [11].

Key features of practical metro DWDM transport systems

System Parameter	Metro core	Metro access
Bit rate	2.5 Gb/s	2.5 Gb/s
Number of channels	40	8-16
Channel spacing	0.8nm	20nm
Maximnum span reach	80 -110 km	up to 15 km
Topologies	Ring, Point-to-Point Mesh,	Ring, Point-to-Point,
Protocols	SONET, ATM,	Mesh SONET, ATM,
IP, ESCON, Fiber	IP, ESCON, Fiber	
	Channel, GBEthemet	
	Channel, GBEthemet	

The Characteristics of metro DWDM transport systems are fundamentally different from long haul DWDM networks. First, the metro DWDM transport system must support the transmission variety of protocols such as SONET, ATM (asynchronous transfer mode), IP (internet protocol), Ethernet, Fiber Channel and ESCON (Enterprise Serial Connection) over different wavelengths which is not the case in a long-haul network. In a long haul network, the traffic usually needs to be converted into standard signal format SONET before the transmission. Second, it is important that the metro transport systems be cost-effective, flexible, scalable and transparent to the bit rate and protocols. Another critical feature is the interoperability, which means the system must work with the legacy equipment such as SONET and ATM technologies. Hence, the mix of a variety of services combined with different protocols makes the metropolitan transport system a challenging environment.

Furthermore, the rings in DWDM metro transport systems must incorporate protection mechanisms to detect failures and reroute the traffic away from the failed nodes or links. A simple ring topology known as unidirectional path-switched ring (UPSR) architecture is depicted in Figure [12].

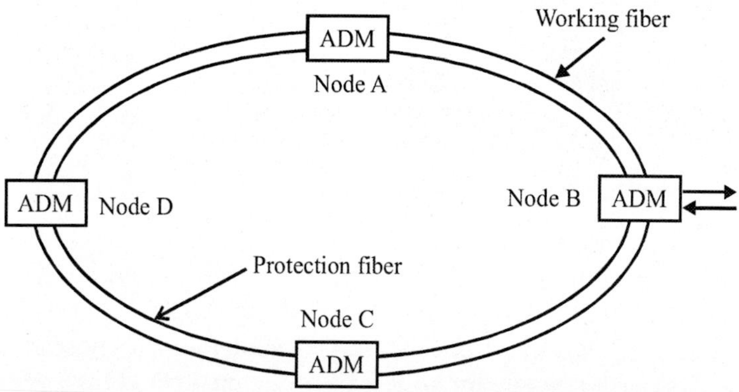

Fig. 4.2 Unidirectional path-switched ring (UPSR) architecture

As can be seen from Figure the UPSR architecture consists of a working fiber and a protection fiber, both' carrying the same traffic along the ring. The working fiber carries the traffic in a clockwise direction, whereas the protection fiber carries the same traffic in the opposite direction, yielding path protection; For example, in the case of fiber cut (working) between nodes A and B, node B is switched to the protection fiber to avoid a network failure. The UPSRs are commonly used in metro access networks, where the traffic from each node is distributed to multiple users. The UPSRs are easy to implement and incur low cost A more efficient use of the fiber bandwidth can be accomplished by using bidirectional line-switched ring architectures (BLSRs) and these ring structures are commonly utilized in metro core or regional networks[12].

The popular topology being considered for the next generation metro networks is the mesh architecture. Adopting a mesh network in place of a ring network provides efficient routing of traffic and bandwidth usage. A typical mesh topology is shown in Figure.

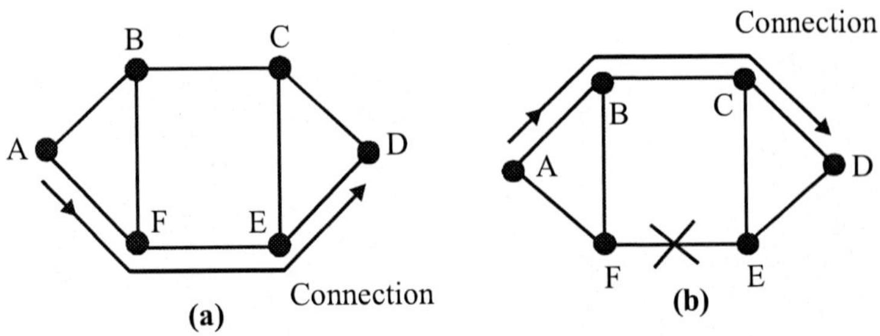

Fig. 4.3 A mesh based metro system architecture; (a) Normal operation; (b) Path-switched operation.

The normal traffic flows between nodes A and D as depicted in Figure, in the case of a fiber cut between nodes E and F, the traffic can be routed through the nodes B and C. The mesh topology offers full recovery from the single point failure. These mesh networks are suitable for wide area or regional networks and can be linked with ring architectures forming hybrid topologies. The mesh networks have high potential to be deployed in future metro networks [12].

4.5 Enabling Technologies

The critical elements which enable an end-to-end DWDM metro transport system operation are transport, switching and network management and control. The transport systems include laser transmitters, receivers, WDM technology, FEC, optical amplification and optical add-drop multiplexers. In the area of switching," an optical switching element should provide transparency to the bit rate and transmission protocols. In addition, efficient management of the network is critical in terms of performance monitoring, reliability and reconfiguration of the network to varying traffic demands.

Our Concerned optical amplifiers serve as the key building blocks of a long-haul link and are also required to metro core networks to overcome the losses of optical nodes and fiber spans. An optical node incurs a large insertion loss due to OADMs, optical switches and MUX/DMUX devices. The initial development of metro core networks utilized EDFAs, but in recent years, semiconductor amplifiers (SOAs), linear optical amplifiers (LOAs) as well as erbium doped waveguide amplifiers (EDWAs) have been proposed for metro core networks[13]. These amplifiers have a smaller footprint and are loss expensive than EDFAs. Additionally, these devices can easily be integrated with other optical devices such as lasers, modulators and multiplexers. Another challenge is that-the amplifiers for CWDM systems should have a modular design to accommodate a wide gain spectrum. Different amplifier design optimizations for the different sub-bands are essential to support the entire spectrum. Here is the point, where the proposed Amplets can be used.

4.6 Introducing the Amplets

New architectures that take advantage of low cost, modest performance amplifiers are being developed today for many different applications. These amplifiers have been called" Amplets". It has reduced performance and lower cost as compared to a traditional broadband WDM amplifiers for example as an against 32 to 40 ITU channels of typical broadband DWDM OA, subbanded DWDM system application amplets are being used with only 4 to 8 ITU channels, therefore total power requirements are reduced by as much as 9 to 10 dB for an EDFA [14] this allows for less pump power and fewer gain stages resulting in a much lower cost product.

In a metro network design, it is important to quantify the number of Amplets required in the network and also determine the exact placement of these amplets [15] For example, consider a given metro ring topology as shown in Figure.

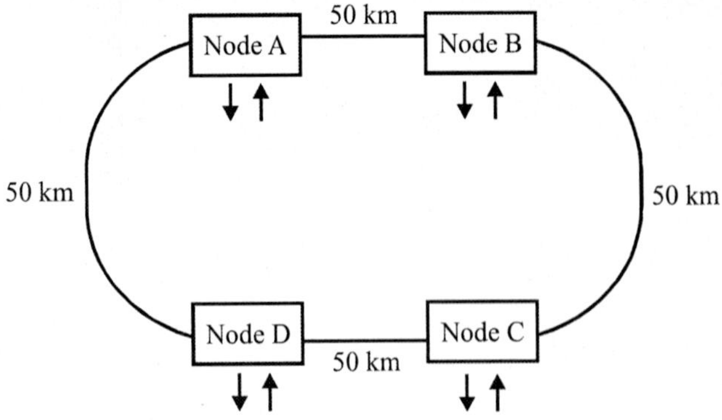

Fig. 4.4 A four node metro ring structure

The ring architecture shown in Figure has a circumference of 200 km and comprises four OADM nodes, each having a loss of 4 dB and a fiber span loss of 10 dB. Assuming a channel power with a receiver sensitivity of -20 dBm and a link margin of 3 dB, the required amount of amplification for the ring should be 41 dB. In this case, two amplets each having a gain of about 20 dB are required to compensate the losses in the ring.

4.7 Design Issues in Amplets

Any optical Amplifier is characterized by parameters like gain, noise figure, bandwidth, gain flatness. The particular choice of amplifier parameters can be changed to optimize the amplifier for a particular application. For example, a system with a long chain of amplifiers requires minimized, noise figure and maximized bandwidth parameters. Noise figure should be minimized because as a signal passes through a chain of amplifiers, the signal is degraded by the addition of noise from each amplifier. This signal to noise degradation will eventually increase to the point where error free transmission can not be obtained.

Minimizing the noise figure minimizes the noise that each amplifier contributes, thus allowing longer chains of amplifiers. The bandwidth should be maximized because passing through a long chain of band- limited amplifiers causes the total end-to-end available bandwidth to be decreased at every amplifier site. That is each amplifier acts some what like a bandpass filter. Those signals on the edge of the pass band will get less gain than those signals in the middle of the band. As the signals pass through more and more amplifiers, the signals in the middle of the amplifier band become enhanced relative to those at the edge of the passband. In a long string of amplifiers, the signals at the edge of the band may become unusable because of this effect.

For amplet applications where-the object is to only increase the output power of a laser so to increase the single span distance available, noise figure is not an important design parameter. In this case only one amplet is used. Therefore, the noise contribution of that one amplifier is not so important. Instead, it would be optimum to maximize the gain and total available output power from the amplifier.

Different amplet application-dictates the required performance and operational characteristics of an amplet but a few critical requirements are shared by all. Amplets are expected to be of low cost. Price is dictated by performance and function but must be sufficiently low to supplant other architectural choices. Amplets must be small in sizes. Functional Packing densities of system will increase over time to lower both first installation and then operational expenses. Amplets must have low Power consumption. As systems get smaller cooling becomes more problematic. Amplets must be easy to use.

4.8 Amplet Design and Technology

There are two fundamentally different amplet technologies that utilize erbium-doped glass as the gain medium:

A. Fiber based EDF A

B. Planar waveguide-based EDWA

C. SOA based amplets

4.9 Fiber Based Erbium Doped Fiber Amplets

Since the first report in 1987, the erbium doped fiber amplifier has revolutionized optical communication. EDFAs provide high gain, high power & low noise figures. .More importantly all the optical channels can be amplified simultaneously within the EDFA in one fiber, thus enabling WDM technology. Tremendous progress has been made in the development of EDFAs, semiconductor pump lasers, passive components and splicing and assembly, technology.

Recent improvements in the control of concentration quenching and reduction in cladding diameters to 80 mm have yielded fibers that are much better suited for amplet use. The gain medium in amplet is specially fabricated fiber whose core is doped with erbium. This fiber is pumped by a semiconductor laser which is coupled by using a wavelength selective coupler. Design choices of pump favour 980 nm devices for their improved noise performance and reduced power consumption.

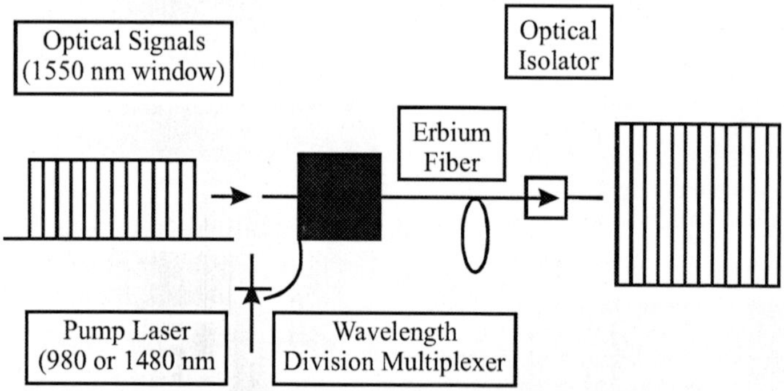

Fig.4.5 Schematic diagram of EDFA

New cooleriess min DIL pumps are currently being offered at upto 200 mw operating power[16]. These devices must operate over large temperature and drive current ranges.

Mostly it is found that pump gain peak shift by greater than 20 nm as the temperature changes from 0 to 70^0 c and the output is varied from 20 to 200 mw. The pumps are wavelength locked with a fiber grating, often with polarization maintaining fiber for improved lock-range over all polarization states so that the pump energy-remains centered in the peak erbium abrorption region. Although the mini DIL pumps are intended to be lower in cost than the larger pumps their small size and reduced power consumption are the primary drivers for use in a EDFA amplet.

New advances in erbium doped fiber have created an opportunity to shrink the package size while still maintaining the performance of larger single pump amplifiers. Newly available fibers have peak absorptions greater than 30 db/m while maintaining pumping efficiency and satisfied noise performance [17]. These fibers allow shortening of fiber by as much as one third as compared to standard EDF optimized for DWDM.

4.10 Erbium Doped Waveguide Amplets

Erbium Doped waveguide Amplifiers differ from EDFA in terms of structure and characteristic. The great advantage of erbium doped planer waveguide technology is the integration of many functions onto an easily produced photonic integrated circuit. In 1994 Hattori et all demonstrated a planar amplifier integrated with a pump-signal combiner having a threshold of 55mw and a gain of 27 dB at 264 mw of pump power [18]. The length of planer amplets is limited by the optical attenuation length of the material to tens of centimeters. To achieve significant gains in distances of order 10 cm requires a concentration of erbium approximately 10-20 times higher than that of EDFA.

At higher concentrations the fraction of closely spaced Er ions increases. Unfortunately at close spacing, Er ion pairs exhibit non radiative up-conversion when both ions have populated upper levels. One electron decays non radiatively to the ground state, transferring its energy to the other electron, which is promoted to a higher level. This problem is accentuated by the clustering of Er ions in silica doped to high Er concentrations. Up-conversion is less pronounced in soda lime glass than is silica becaue the cage like structure soda lime glass tends to separate the Er ions and reduce clustering. Therefore a core of soda lime glass is proposed for EDW Amplets. In 1996 ghosh et all have demonstrated an amplification with a threshold of 8 mw and a 4-dB gain at 80 mw in a 6 cm. long amplifier [19].

4.11 SOA Based Amplets

SOA device design is similar to semiconductor laser design. The typical SOA is an MOCVD-grown layer structure consisting of an active layer sandwiched between p and n doped cladding layers which allow current injection. Lateral optical confinement is accomplished by etching a mesa, which is overgrown with a current blocking structure, which can be semi insulating indium phosphate or a diode structure in reverse direction as shown in Fig. below

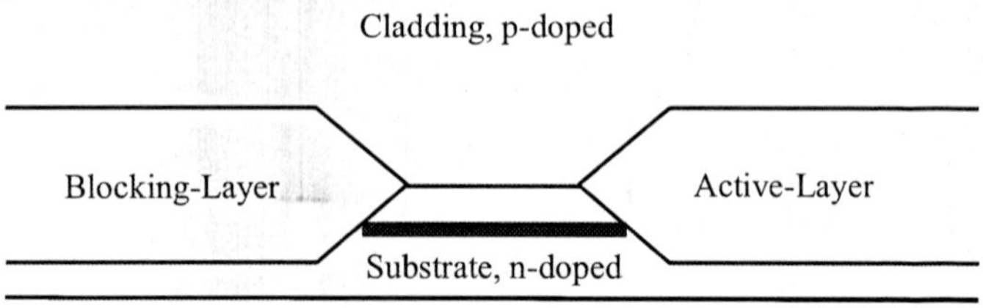

Fig.4.6 SoA Device Structure [21]

A major advantage of SOA is its larger gain band width. The gain spectrum of SOA is intrinsically flat and wide (80 nm). With the combination of several SOAs centered at different wavelengths, the whole transmission band of optical fiber can be covered. Another key advantage of SOA based amplet is the potentially low manufacture cost. SOA with multiple gain stages can be integrated on the same chip during fabrication. Small size and low power dissipation are additional advantage that make it suitable for access network [20].

4.12 Comparing Different Amplets

A comparison table is presented below bassed on output power gain, noise figure, size, power consumption [21].

	EDFA Amplet	**EDWA**	**SoA**	**Unit**
Output power	> 15	> 12	> 13	dBm
Gain	> 24	> 20	> 25	dB
Noise Figure	< 6	< 6	< 8	dB
Max gain Flatness Deviation	< 0.75	< 0.75	< 0.1	dB/nm
Size	45′70′12	70′18′12	30′12.5′12.5	mm
Power Consumption	< 1	< 1	< 5	w

Based on above parameters we can select the amplet to suit our need.

4.13 Conclusion

In this Chapter different types of amplets are proposed and their characteristics have been studied for metro network. These amplets have been found to be cost effective and compact to suit the need of present day service providers. As more and more number of internet connection are coming up and more cities are getting computerised the amplets will become a necessity to exploit the data speed and band width of optical fiber.

References

[1] Yongg yoo kim and Jichai Jeong; "Transmission performance of 10-Gb/s 1550nm transmitters using semiconductor optical amplifiers as booster amplifiers". JLT, Vol. 21, No. 2, Feb. 2003, pp 476-495

[2] W.Choi, S Hur et all; "Transmission performance analysis of 8x10 Gbps WDM signals using cascaded SOA", JLT, Vol. 20, Aug. 2002, pp 1350-1356

[3] LH Spiekman, JM Wieenfeld; 8x10 Gbps DWDM transmission over 240km of standard fiber using a cascade of SOA"; IEEE Photon. Technology letter, Vol. 12, Aug. 2000, pp 1080-1084

[4] Thomas Metal; "System advantages of tunable lasers in the metro DWDM environment"; National fiber optic Engineer Conference proceedings, 2001, pp. 1486-1866

[5] JJJ Crijns, LH Spiekman et all; "Static and dynamic switching performance of a metro WDM ring using linear optical Amplifiers;" IEEE Photonics Technology letter, Vol. 14, Oct. 2002, pp 1481-1483

[6] Kaminow, I.P. et all; "A wideband all-optial WDM networks"; IEEE Journal on selected area in communications, Vol. 14, No. 6, 1996, pp. 780-799

[7] Hall. E. et all; "The rainbow-II gigabit optical network"; IEEE Journal on selected area in communications, Vol. 14, No. 6, 1996, pp. 814-823

[8] Saleh AM et all; "Architectural principles of optical regional and access networks"; Journal of lightwave technology, Vol. 17, No. 12, 1999, pp. 2431-2448

[9] Wanger, R.et alll; Multiwavelength optical Networking"; JLT, Vol. 14., No. 6, 1996, pp. 1349-1355

[10] Tamkos I. et all; "Demonstration of negative dispersion fibers for DWDM metropolitan area networks"; IEEE Journal on selected area in communications, Vol. 7, No. 3, 2001, pp. 439-460

[11] Internet data/online/available; 2008, http//www.eiena.com/

[12] antoniades N. et al; "Performance and topological design of metro WDM networks using computer simulation"; JLT, Vol. 20., No. 1, 2002 , pp. 149-165

[13] Gerlas Van Den Hoven et al; "Semiconductor amplifiers meet metro requirements"; WDM solutions; October 2001, pp. 45-54

[14] AK Srivastava, Y Sun et all; "EDFA transient responses to channel loss in WDM transmission system"; IEEE Photonics Technology letter, March 1997, pp 336-388

[15] Rama Murthy B "Design of optical WDM Networks;" LAN, MAN, WAN Architetures Academic Publishers 2001, pp. 181-197

[16] S Mohedic, T Pliiska et all; "Coolerless operation of 980 nm pump modules"; Opticial Fiber communicaton conference, 2001, Vol. 3, pp WDD 77-1 – WDD 77-3

[17] S. Tammela, M. Hotoleanu et all"; Very Short Erbium doped Silica Glass fiber for L-Band Amplifier:' Optical Fiber Communication Conference, 2003, pp. 376-377

[18] Hattori, KT Kitagawa et all;" Erbium doped Silica based waveguide amplifier intergrated with a 980/1350 nm WDM Coupler"; Electron Letters, Vol. 30, 856-857

[19] Ghosh RN, J. Schmulovich et all"; 8 mw threshold Erbium doped planer waveguide amplifier"; IEEE Photonics Technology letter, Vol.-8, pp 518-520

[20] AK Srivastava, Y Sun at all; "Optical Amplifiers for Terabit Capacity Light wave system"; WDM Technology, Academic Press 2005, pp 321-367

[21] Donald R Zimmerman; "Ampliffier for the Masses"; JLT, Vol. 22, No. 1, Jan 2004, pp. 63-70

5

Integrated Optical Amplifiers

5.1 Introduction

The operating speed of lightwave communication systems has increased dramatically over the last several years. Currently, research is focused on 40 Gb/s systems and beyond at many research and development laboratories. The integration of optical components or devices plays a key role in the' design of these high speed lightwave systems in terms of cost savings and complexity. [1] The development of EDFA in optical fiber is a revolutionary step in optical communication.

These amplifier can also be made as OICs. As the performance of planer amplifiers improves and hybrid-optical integration makes it possible to assemble an entire amplifier consisting of a semiconductor pump laser a WDM combining pump and signal wavelengths, an amplifying waveguide and other components in a single compact package, this amplifier will become attractive and competitive with the optical fiber amplifier for some applications. Waveguide amplification is the fundamental of Integrated optical amplifiers.

5.2 Waveguide Technology

The most fundamental building block of an integrated optic circuit is waveguide. There are several ways by which waveguides and devices in the IO form can be realised. The type of material chosen more or less decides the process technology to be employed.

In the case of glass, wet and dry ion exchange techniques [2] are commonly used for fabricating mostly passive IO components such as splitters/ combiners. Polymer waveguides on glass and other substrates, on the other hand, are formed by spin or dip coating. While this process is simple, precise thickness and uniformity control are difficult. Plasma polymerisation and Langmuir 'Blodgett method of formation are other techniques used in the case of polymers. The most popular material for IO, $LiNbO_3$, can be processed using either metal in-diffusion (usually titanium) or by proton exchange in weak acids [3] Epitaxial methods (LPE, MBE, MOCVD) are appropriate for the growth of crystalline layers and quantum well structures in semiconductors [4]. Amorphous layers on silicon are also useful in passive component development such as grating and channel waveguides for WDM and LAN applications.

There are certain pre-fabrication procedures which are to be carefully followed before an actual waveguide/device process is initiated. This is absolutely essential if low loss IO components and systems are to be achieved. Substrate preparation in terms of polishing and cleaning are the critical pre-fabrication processes. Dust free environment, temperature and humidity control and prevention of inadvertent contamination are additional precautions to be followed.

5.3 Integrated Erbium Doped Waveguide Amplifier

The great promise of erbium-doped planar waveguide technology is the integration of many functions onto an easily mass-produced photonic IC. Great strides have been made in this area with recent results yielding amplifiers of, very good performance with high degrees of integration. Two basic technologies have been used to achieve these recent results: plasma-enhanced chemical vapor deposition (PECVD) and metal ion exchange (IE). With PECVD the passive and erbium-doped waveguides are deposited directly onto the same silicon substrate in an integrated fashion [5]. With IE technology, metal ions are embedded into a glass substrate to selectively raise the refractive index in the waveguide. Erbium-doped glass is used for the active waveguides and clear glass is used for the passive waveguides. The active and passive sections are then joined together for the final integration.

The integration of all necessary amplifier components is hampered by the availability of Faraday-effect materials for integrated isolation. Most suppliers are currently experimenting with methods to attach bulk isolators to their waveguides with sufficient performance and stability to eliminate fiber coupled devices. The focus of this activity is cost and size reduction.

As with the EDFA, the cost of the pump is a major concern. Maybe even more so since the pumping efficiency is lower in EDWA. One method under consideration for controlling cost and reducing size is to couple a pump directly to the waveguide without an intermediary fiber. The difficulties of achieving a stable and robust pump package are well known and as such we might expect this activity to take some time to achieve commercial acceptance.

5.4 Integrated Amplifiers and Photonic Integrated Circuits

There have been a significant number of development in the technology integration of semiconductor lasers, amplifiers, and other related devices on the same chip. These Chips allow higher levels of unctionality than that achieved using single devices. For example, lasers and electronic drive circuits have been integrated, serving as simple monolithic transmitter. The name photonic integrated circuit (PIC) is generally used when all the integrated components are photonic devices, e.g., lasers, detectors, amplifiers, modulators and complier.

5.5 MQW Amplifiers

For Amplifier applications where the amplifier is integrated with another photonic device such as a laser it is not necessary to have polarization independent gain. Because. multiquantum well (MQW) materials generally have higher gain than regular double heterostructure material, MQW active regions are generally used for amplifiers that are integrated in a photonic circuit. The saturation output power — the power at which gain decreases by 3 dB from its value at low power — is higher for MQW amplifiers.

Semiconductor amplifiers are often integrated with tunable laser to increase the output power. The schematic of such an integrated couples device is shown in Fig. The waveguide layer with a band gap of ~1.1mm couples the laser and the amplifier devices. For the structure shown both the laser and the amplifier have a MQW active region.

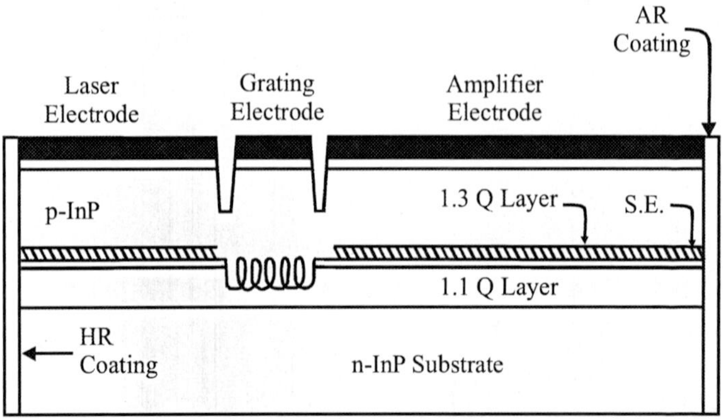

Fig 5.1 Schematic of a photonic integrated circuit with a DBR laser and amplifier.

5.6 Wideband Hybrid Amplifiers

In recent years, the rapid growth of internet data traffic has created a greater need for wider bandwidth lightwave transmission systems. Many WDM transmission experiments have achieved system capacities of the order of several Tb/s using several types of wideband amplifiers [6]. A wideband amplifier that utilizes the dual arm configuration and provides a bandwidth of 84 nm is shown in Figure.

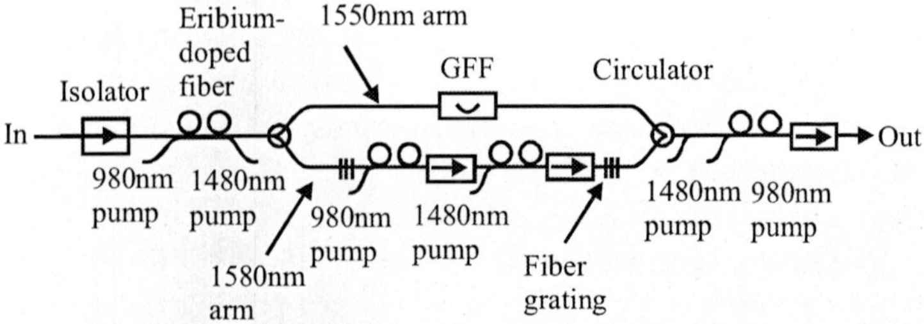

Fig. 5.2 Dual arm gain flattened EDFA for both C and L bands.

The advantage of a dual arm configuration is that each arm can be independently optimized for each band of operation. The lower arm contains a two stage amplifier and is optimized for L band operation, whereas the upper arm contains a gain flattening or equalizing filter, that optimizes the C band amplification. The C and L band signals are recombined after the two arms. The total pump power from all the sources required for high gain operation is of the order of 500 mW.

5.7 Multichannel WDM Sources With Amplifiers

An alternative to single-channel very high speed (>20 Obis) data transmission for increasing transmission capacity is multi channel transmission using wavelength division multiplexing (WDM) technology. In a WDM system many (4, 8, 16 or 32) wavelengths carrying data are optically multiplexed and simultaneously transmitted through a single fiber. The received signal with many wavelengths is optically demultiplexed into separate channels, which are then processed electronically in a conventional form. Such a WDM system needs transmitters with many lasers at specific wavelengths. It is desirable to have all of these laser sources on a single chip for compactness and ease of fabrication, as in electronic integrated circuits.

5.8 Spot Size Conversion (SSC)

A typical amplifier diode has too wide a beam pattern (~30° x 40^0) for good mode matching to a single-mode fiber. This results in a loss of power coupled to the fiber. Thus, an amplifier whose output spot size is expanded to match an optical fiber is an attractive device for low loss coupling to the fiber without a lens, and for wide alignment tolerances. Several researchers have reported such devices using lasers [7]. Generally they involve producing a vertically and laterally tapered waveguide near the output facet of the laser. The tapering needs to be done in an adiabatic fashion so as to reduce the scattering losses. The schematic of a SSC laser is shown in Fig. The amplifier designs are similar. The laser is fabricated using two MOCVD growth steps. The ssc section is about 200 mm long.

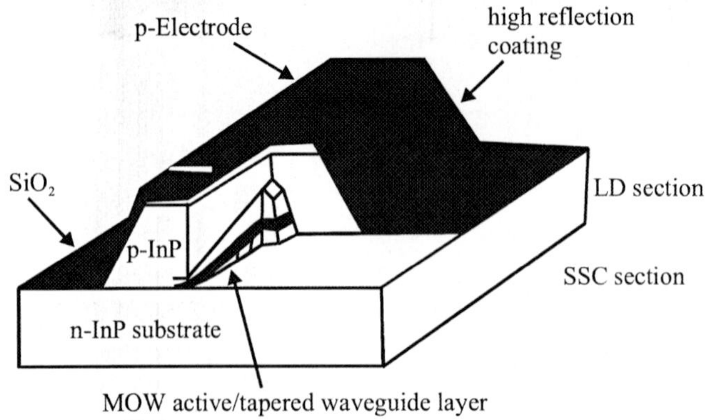

Fig 5.3 Schematic of a spot size converter (SSC) laser

The waveguide thickness is narrowed along the cavity from 300 nm in the active region to ~100 nm in the region over the length of the SSC section. The laser emits near 1.3 mm, and pas a multiquantum well active region and a laser section length of 300 mm. Beam divergences of 9^0 and 10^0 in the horizontal and vertical direction have been reported for SSC devices.

5.9 Mach-Zehnder Interferometer

Semiconductor optical amplifiers incorporated into Mach-Zehnder interferometer (MZI) or Michelsen interferometer configurations have been used for wavelength conversion and to demultiplex high-speed time division multiplexed optical signals. The schematic of the device is shown in Fig. The input signal beam is split into two beams by an input y-junction, which propagate through semiconductor amplifiers positioned in the upper and lower arms of the interferometer. These beams then merge and interfere at an output y-junction. The phase of the interference is altered by injecting another signal in one of the other ports.

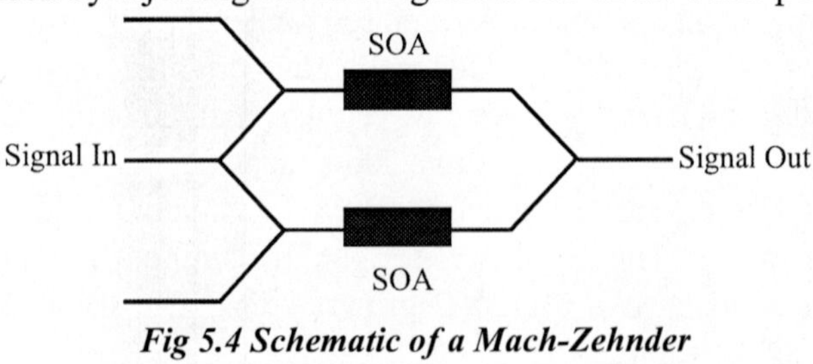

Fig 5.4 Schematic of a Mach-Zehnder

The ratio of the output (P_0) to the input signal (P_t) is given by

$$\frac{P_0}{P_i} = \frac{1}{8}\left\{G_1 - G_2 - 2\sqrt{G_1 G_2}\cos(\phi_1 - \phi_2)\right\} \text{ [8]}$$

Where G_1, G_2 are the gains of the amplifiers and ϕ_1, ϕ_2 are phase changes induced by nonlinear effects in the two amplifiers. The phase difference, $\Delta\phi = \phi_1 - \phi_2$, at the output is given by

$$\Delta\phi = \frac{\alpha}{2P_s}\left\{-\frac{hv}{e}(I_1 - I_2) + \frac{P_c}{2}(G_1 - 1)\right\} \text{ [8]}$$

where $\alpha = \delta n_R / \delta n_{IM}$ is the ratio of the carrier-induced change in the real and imaginary parts of the index, also known as the line width enhancement ft. factor. P_c is the injected power in one of the amplifiers. P_s is the saturation power in the amplifier. and I_1, I_2 are the currents through the two amplifiers. For $I_1 \cong I_2$ and $G_1 \cong G_2$, Eq. reduces to

$$\frac{P_c}{P_i} = \frac{G}{2}\sin^2\left(\frac{\Delta\phi}{2}\right) \text{ [8]}$$

Where $\Delta\phi = \frac{\alpha}{4}\frac{P_c}{P_s}(G - 1)$

Using typical values $\alpha = 3$, $P_s = 10$ mW, and experimental saturated gains $G_1 \cong 40$, we get $P_c \cong 1$ mW for $\Delta\phi = \pi$ Thus. the optical power that needs to be injected in one of the other ports for π phase shift can be easily achieved. There may be some thermally induced refractive index changes but these are considerably smaller than the carrier-induced changes previously considered and can be ignored.

5.10 Conclusion

Integrated optical Amplifiers have been remarkable range of capabilities. Its ability to be integrated with other communication devices have proved to be more fruitful in photonic revolution. Wave guide technology has been discussed and found that coupling and propagation losses have been reduced to the point that the fiber-waveguide-fiber insertion loss of a 6 cm. straight waveguide is as low as 0.3 dB. The capability to integrate many SOAs on the same chip with inter connecting waveguides have been discussed in hybrid integration. The need for higher capacity is continuing to the development of semiconductor amplifiers as Mach-Zehnder and as low power amplifiers in integrated devices. Recent Development in Spot Size converter integrated amplifiers is quite progressive but some more work needs to be done to make their manufacturing easy.

References

[1] Inchino H. et all; "Over 10 Gb/S IC's for future lightwave communication"'; JLT, Vol.-12, No. 2, 1994, pp 22-28

[2] Ramaswamy R and R. Srivastava; "Ion Exchange glass waveguides"; JLT, Vol.-6, 1994, pp 984-1002,

[3] Loni A, G Hay et all; "Proton exchanged $LinbO_3$ waveguide"; JLT, Vol.-7, 1989, pp 911-919

[4] Jeri RJ and K Kapor; "Low loss semiconductor optical waveguide"; IEEE Journal of Quantum Electronics, Vol.-27, 1991, pp 626-640

[5] K. Shuto; K Hattori et all; "Erbium doped phosphosilicate galss waveguide amplifier fabricated by PECVD"; Electron letters, Vol.-29, No. 2, 1993, pp 139-141, Jan 1993

[6] Masuda H, "Review of wideband hybrid Amplifiers"; optical fiber communication conference, optical society of America, 2000, pp 198-200

[7] RY Fang, D. Bertone, M Meliga, I Montrosset; "Spot Size converter integrated amplifiers" Photonics Technology letters, 9 (1997), pp. 1084-1086

[8] SL Danielsen, Ps Hansen et all; "MZI with integrated SoAs"; IEEE journal of quantum Electronics, 16 (1998), pp 2095-2108.

6

Fabrication of a Novel Integrated Amplifier : Optical Microring Amplifier

6.1 Introduction

For more than 30 years, optical communication and integrated optics have been in the limelight of both academic as well as industrial research. The dramatic reduction in transmission loss of optical fiber coupled with important developments in light sources, detectors and controllers has brought about a phenomenal growth in optical communication and networks. The very large bandwidth, of the order of 10^{15}Hz, gives optics a potential advantage for signal processing which is well beyond the capability of electronics. Some of the main advantages which made optical communication to replace conventional networks are low attenuation, high bandwidth, small physical size and weight, immunity from interference and crosstalk, signal security, ruggedness, flexibility, reliability and cost effectiveness. As integrated optics or photonics, the optical counterpart of electronics developed, several other attributes have also became apparent such as optical signal processing and computing. Wavelength division multiplexing (WDM) has been recognized for a number of years as a promising and applied technology that can be used to increase the aggregate throughput of the optical fiber. Recently, more attention has been given to devices and technologies that can be exploited to enable WDM to move from being a pure transmission technology into a state where it will be applied in all optical network. In particular, devices and technologies such as wavelength routed switches, optical cross connects, optical add/drop multiplexers, switched sources, tunable sources, tunable filters, wavelength converters etc.

Large scale integration of photonic components is the leading target of modern optical technologies, which are oriented in the direction of an increasingly higher number of optical functions per manufactured chip, in order to improve both complexity and cost effectiveness of components [1].

In the last two decades several photonic integrated circuits (PIC) have been developed such as, amplitude and intensity modulators, tunable wavelength filters, frequency shifters, phase modulators, polarization controllers, spatial switches etc. Recently, researchers at Intel claim the world's fastest silicon modulator that could encode data at a rate of 30 Gb/sec, an advance that could cut bandwidth costs and make computer run faster and cooler [2]. Infinera is developing a chip which is capable of sending/receiving a 1.6 terabit information per second on Indium Phosphide [3]. Researches from MIT have developed micro-photonic device which makes all incoming light on single polarization before the data is processed [4]. Amplification at this level is more challenging therefore it calls for newer methods and technology of amplification to come up to meet the necessity. Optical Micro Ring Amplifiers can be the answer to all the Challenges.

6.2 Working Principle of Optical Microring Resonators

Mirco Ring Resonator works on the principle of evanescent field coupling. Typically a resonator is a device, bounded at least in part by highly reflecting surfaces, in which light of particularly discrete frequencies can set up standing wave mode of low loss. Evanescent field can be seen before the critical angle for total internal reflection is reached. In order to understand this phenomena it is necessary to use the wave theory model for total internal reflection at a planar interface. This is illustrated in the figure below, where the arrowed lines represent wave propagation vectors and a component of the wave energy is shown to be transmitted through the interface into the cladding. Following symbols have been used.

A, B, C = Incident, Transmitted & reflected EM wave amplitude

A_0, B_0, C_0 = Wave amplitude Maxima

b = Wave propagation constant

bx_1 = Propagation constant in the x-direction for guide

bx_2 = Propagation constant in the x-direction for cladding

k = Wave Propagation constant in free space.

So, when there is only partial reflection, the field in the cladding is given by

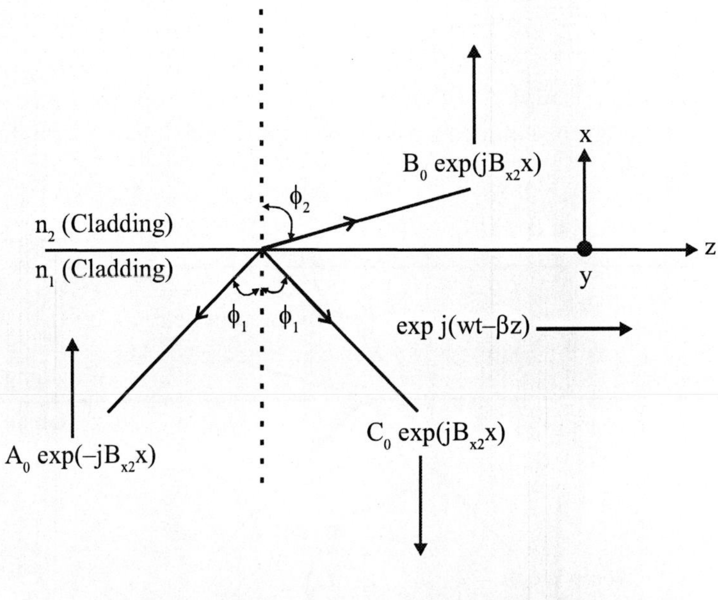

Fig 6.1 Wave Incident on the intterface Showing incident, transmitted and reflected waves

$$B = B_0 \exp-\left(-j\beta_{x2}x\right)\exp j\left(wt-\beta z\right) \text{ ....[2]}$$

Where wave propagation is considered in the Z-direction and β_{x1} and β_{x2} are propagation in the x-direction for the guide cladding respectively. However when total internal reflection occurs, β_{x2} becomes imaginary and may be written as $-j\xi_2$. So the above equation becomes

$$B = B_0 \exp(-\xi_{2x}) \exp j(wt - \beta z)$$

This equation gives the transmitted wave in the cladding which indicates that amplitude of the field in the cladding is decaying exponentially. Such a field exhibting an exponentially decaying amplitude is referred as Evanescent Field. A field of this type stores energy and transports it in the direction of propagation along with the field in the core.

An optical ring resonator consists of a single mode channel waveguide in a ring form with another straight waveguide in close proximity which is often referred as bus wave guide and hence evanescently coupled, as shown in Figure. This configuration is generally referred to a 'notch' filter. Light from the straight waveguide is coupled to the ring resonator and propagates along the circumference of the ring and re-couples back into the straight waveguide.

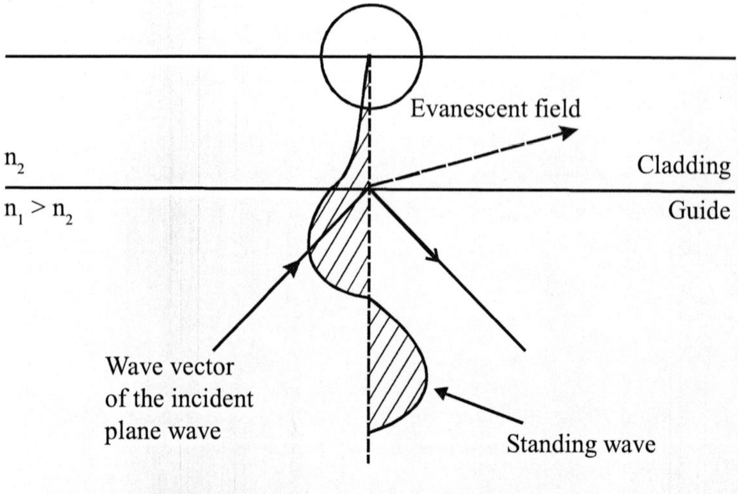

Fig 6.2 Evanescently Coupled Microring

6.3 Optical Microring Resonators Analysis

Analysis of ring resonant structures dates back to 1910 when lord Rayleigh first proposed it to the field of acoustics later, in 1960s, it was further developed for microring applications along with masers. In 1970s ring resonators were studied for suitability in optical domain. Ring resonators have been found to enhance the phase changes in the signal greatly in near resonant wavelength.. This has been used for many applications such as pressure and vibration sensing, resonant routing, delay lines, logic gates and all optical processing. Fabrication of resonators with ring length 2-100 microns with waveguide dimensions of about 1.5mm x 1.5mm have been reported. They are found to be integrated closely to have densities of 10^5 devices/cm^2.

Here in this, I propose a different application of optical amplification in ring resonator. By suitably doping the ring and confining the coupled wavelength to undergo required number of rounds an amplification of a long fiber waveguide equivalent can be obtained within a small structure of ring resonator. This requires analysing a plane undoped resonator in terms of its output characteristics, field en hancements, free spectral range, finesse etc. which are accountable for lossless propagation of wavelength in ring structure.

Optical micro-resonators are at present considered as basic elements for large scale integrated optics typically for applications in WDM and DWDM [5]. These devices possess the desired characteristics, flat in band response and high rejection in the stop band, for filtering, packing, separating, routing and managing large numbers of closely spaced channels and could be the key elements to increase the spectral efficiency of WD M systems [6]. One reason for the lack of systematic large-scale photonic integration is that, until recently, there has been no universal building block for integrated optics comparable in functionality to the transistor for integrated electronics. This is beginning to change, however, as researchers worldwide demonstrate high-quality optical signal processing devices that are several orders of magnitude smaller than conventional devices. Central to this miniaturization is the micro-cavity resonator, a versatile elemental building block suitable for massive integration [7].

Optical micro-cavities confine light to small volumes by resonant recirculation. Figure shows a couple of structure of integrated optical resonant micro cavities.

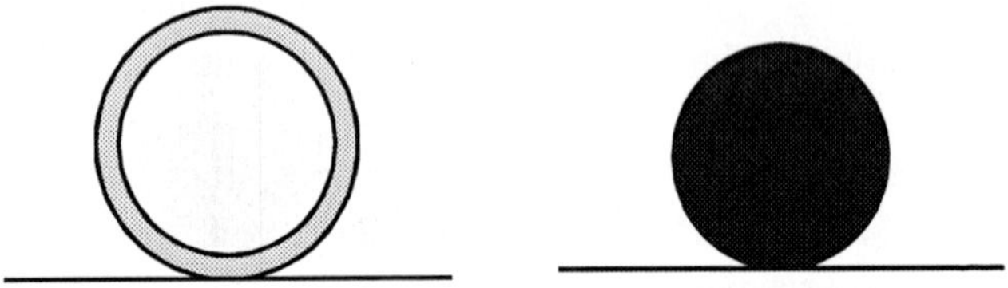

(a) Ring Resonator (b) Disk Resonator

Fig 6.3 (a) Ring and (b) Disk Resonator for Micro Cavities

Both the ring and disk structures have been studied in great detail. Disk resonator has the advantages of low loss and low line-width around resonant wavelength when compared to ring structure. However, it has a serious drawback of being mostly multimoded limiting its usefulness. Devices based on optical micro-cavities are already becoming indispensable for a wide range of applications [8].

6.4 Applications of Optical Micro-Cavities

Small size of micro-cavities are suitable for Cavity Quantum Electro Dynamics (CQED) [9] since the Whispering Gallery modes can be made as small as a few 10-100 cubic wavelengths and the electric field amplitude per photon becomes sufficiently high to realize CQED and for manipulating quantum entanglement with atoms and photons. Micro lasing properties have also been studied in neodymium-doped silica micro-spheres [10]. A new generation of optical microresonators is making possible the exploration of quantum electrodynamic phenomena in condensed matter .systems and providing microlasers with a wide range of potential applications

There are many other fields and many topics, such as soliton effects effects in quantum-well micro-cavities [11], enhancement and suppression of spontaneous emission, bio-sensing etc. In quantum optical devices, micro-cavities can coax atoms or quantum dots to emit spontaneous photons in a desired direction or can provide an environment where dissipative mechanisms such as spontaneous emission are overcome so that quantum entanglement of radiation and matter is possible.

Integrated optical micro-resonators have two important attributes that make them a key new technology for future optical communications, namely, multi-functionality and compactness. Functionality refers to the fact that micro-resonators can provide a wide range of optical signal processing functions, including channel-dropping filters [12]. WDM demultiplexers [13], true ON OFF switches [14], dispersion compensator's [15]. Rotation sensors polarization rotators, wavelength converters and greatly enhanced nonlinear optical effects to name just a few. Further, we can systematically improve the response of resonator-based devices by implementing multiple-cavity arrangements [16]. Micro-resonators can be made ultra-compact since devices having dimensions down to the scale of optical wavelength have been proved. Integration densities of up to 10^5 devices/cm^2 can now be envisioned [17].

The following is a list of futuristic applications, to name a few, in which optical micro ring resonators have found a key place

Cavity quantum electrodynamics	Interferometric Sensing
High resolution spectroscopy	Pressure and vibration sensing
Disk laser	Dispersion compensation
Laser frequency stabilization	Resonant routing
Bistable optical elements	All optical processing
optical memory	Logic gates
ON/OFF switches	Channel dropping filters
WDM multiplexers & demultiplexers	Polarization rotators
Wavelength conversion	Delay lines

6.5 Reviewing Micro Ring Resonators

In this section an overview of current scenario regarding optical micro ring resonators in communication and networking has been presented. Optical filters with optimum bandpass designs using ring resonators have proved requiring fewer couplers and phase shifters as compared to a general waveguide based integrated optic structures. Also, they have been robust to loss. A 200-GHz FSR periodic mux/demux with flattened transmission and rejection band of 75 GHz 1 dB bandwidth has been fabricated using a Mach-Zehnder interferometer along with an IO-micro ring resonator [13]. An optical micro-ring filter with curved sections of radius 2.25mm and straight sections of length 1 mm has been reported with 0.8 nm(100 GHz) of tuning using electro-optic effect on GaInAsP-InP [18] A recent study on polarization rotation effects in optical ring resonator based devices showed that polarization conversion is enhanced by the cavity effect and becomes strongly frequency dependent so that the spectral response can be highly distorted even at a low strength of polarization coupling. The high Q-factor associated with the whispering gallery modes at the surface of silica micro-sphere is utilized for enhanced sensitivity in aqueous mercuric ion detection using microchip optical ring resonator sensors. High confinement suspended micro-ring resonator in silicon-on-insulator was demonstrated with Q-factor exceeding 15,000 and insertion loss less than 0.2dB per suspension bridge [19]. An IO micro ring resonator displacement sensor has been proposed for scanning probe micros copies using GaAs-based cantilever with integrated ring and race-track resonators Analytical models were pursued, which allow a direct comparison of the different all optical switch geometries leading to optimization of ultrafast all-optical resonator switching [20]. Using nano-imprint technique, polymethylmethacrylate, polystyrene and polycarbonate micro ring resonators were fabricated and were shown to give Q factor 5800. A free-standing all optical polymer micro ring resonator optical filter is demonstrated with extinction depth measured to be -27dB of the maximum output signal [21]. Waveguide coupled AlGaAs/GaAs micro cavity ring and disk resonators with high finesse of 120, Q factor greater-than 8500 and 21.6 mm FSR are reported. Tunability is achieved based on the interaction of a micro actuated dielectric/metal membrane over an Si_3N_4 micro-ring waveguide giving tuning range of 27 nm. Micro ring resonators were also fabricated on silica-on-silicon technology using deep glass etching with ring radius of 25 mm [22]. Optical Sensing of biomolecules using micro-ring resonators has also been demonstrated utilizing avidin-biotin binding chemistry. Second and third order serial-cascaded racetrack filters in Silicon-on-Insulator(SOI) have been reported and are found to have flatter passband and steeper roll-off over single track resonator in addition to little polarization dependence SiO_xN_y technology has been used for design and characterization of ring-resonator based filters with free spectral ranges of 100 to 50 GHz [23]. A technique for detecting heavy water (D_2O) that is 30 times more sensitive than the best competing approach is demonstrated using a tiny optical micro-resonator from silica and immersing it in varying concentrations of D_2O in H_2O. Because H_2O has greater optical absorption than D_2O, the cavity had a lower Q in solutions with a higher concentration of H_2O. Using the method, the scientists have detected D_2O concentrations as low as 10^{-4} percent [24].

6.6 Fabricating Microring Structure

In 1971, Weber and Ulrich first demonstrated a ring resonator in 5mm glass rod coated with Rhodamine -6G-doped polyurethane [25]. This was followed by Haavisto and Pajer in 1980 with doped poly (methyl methacrylate) film on quartz substrate [26]. In the next decade many groups demonstrated ring resonators using glass modified by compounds like $AgNO_3$, KNO_3, TiO_2 etc.

Several other material systems are now being investigated for micro-ring and microphotonic platforms. $LiNbO_3$ based structures using both Ti-exchange and proton exchange have been demonstrated. Group IV based ring structures being extensively studied starting from their first demonstration of GaAS-AlGaAs based ring resonators by Rafizadeh et al. in 1997 Polymers have also been explored for both passive as well as active micro ring structures. The success of these material systems depends on the extent to which the cavities can be made low loss. Losses are attributed mainly to waveguide sidewall roughness scattering. The large core-cladding index contrasts (typically 1.5:1 up to 3.:1) result in considerable perturbation-induced scattering unless the sidewalls are smooth.

High-index-contrast waveguides used for micro-ring resonators support tightly confined modes, that is, modes whose energy is contained almost entirely within the waveguide core. These waveguides need to be brought into close proximity for mutual interaction. One can have two types of coupling systems. One is lateral coupling in which the bus waveguide and ring are in the same layer while the other is vertical coupling in which the ring and bus waveguide will be in different layers one over the other. The following Figure shows the two coupling systems.

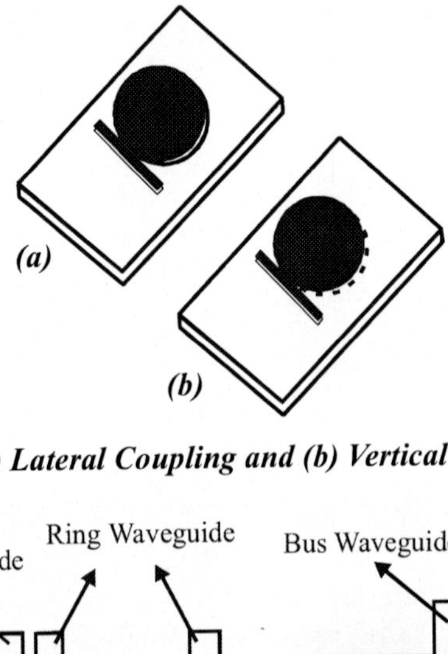

Fig 6.4 (a) Lateral Coupling and (b) Vertical Coupling Systems

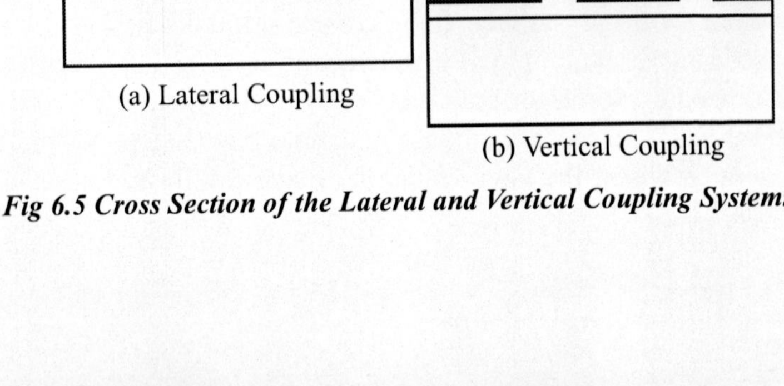

Fig 6.5 Cross Section of the Lateral and Vertical Coupling Systems

Typical cross section of the lateral and vertical coupling systems is shown in Fig. In laterally coupled semiconductor ring resonators where the ring and the bus guides are in the same plane, waveguide separations of 100-200 nm are required. These dimensions call for high-resolution e-beam lithography for defining narrow gaps with high integrity and repeatability. Achieving the required gap spacing is perhaps the most challenging aspect of laterally coupled resonators. Vertical coupling eliminates the need for e-beam lithography and considerably relaxes fabrication tolerances. In the vertical coupling scheme, the coupling between the microring waveguide and the bus waveguide is in the vertical direction through an epitaxially-grown mid layer in contrast to the lateral scheme in which coupling is through an etched low refractive index gap. Also, it gives active-passive integration feasibility.

6.7 Fabricating Microring Amplifiers

The SOL-GEL process is a very simple and effective method for fabricating integrated optical devices where films with varying composition can be easily prepared. Combined with lithographic process, this process can be used to realize several configurations of integrated optical devices. One of the advantages of the sol-gel process that can be effectively exploited is the ease with which dopants can be introduced. In particular, this can be employed to dope active materials such as neodymium, erbium, and cerium to fabricate integrated optical (IO) devices with amplifying and photosensitive properties. The incorporation of active material, such as neodymium (Nd^{3+}) and erbium (Er^{3+}) in host media has potential applications in lasers and amplifiers, while cerium (Ce^{3+}) doping results in better photosensitivity.

Well known glass waveguide passisve device such as star coupler/power splitters, could be suitable modified into amplified version by the incorporation of a rare earth, which thereby would be useful in compensating the splitting and insertion losses. Using rare earth doped waveguides it is also Possible to fabricate monolithic and compact narrow line single-frequency lasers by integrating coupled cavities, resonators, or distributed feedback mirrors. Such potentially stable laser sources would also find application-at 1050 nm (neodymium), suitable for seeding high-power glass laser. Finally, by employing various rare-earth ion transitions, it is possible to reach other wavelengths from the IR to the visible region of the electromagnetic spectrum by suitable up-conversion pumping schemes, as in the case of fiber lasers [27].

6.8 Sol-gel Process

Sol-gel processes are useful for making articles of synthetic silica glass having shapes close to that desired for the final product. This reduces costs associated with machining and polishing the articles. Typically, sol-gel processes involve steps of (1) synthesizing a liquid silica solution, or sol. typically by hydrolyzing an alkoxide precursor or stabilizing a colloidal silica solution, (2) condensing the Sol to yield a wet gel. (3) aging the wet gel to strengthen it, (4) drying the wet gel to produce a dry gel, and (5) sintering the dry gel to yield a dense glass article.

The gels yielded by this sol gel route typically are quite porous, with large pore (void) volumes and relatively low volume loading of silica. Because the porous gels are later sintered into dense glass articles, the large void volumes bring about a large amount of shrinkage during the sintering step and thus yield silica articles of reduced size. It therefore follows that reducing the void volume will increase the amount of sintered silica glass that can be obtained from a single processing cycle and thus reduce manufacturing costs. Increasing the loading of silica in the sol is one way to achieve this[28].

Synthesis of several oxide glasses and waveguides have been known to employ the sol-gel process. Of these, in terms of potential use in fiber optic communications, three compositional group are of importance: (1) titania-silica glasses for passive device (2) rare-earth doped silica glasses for laser and amplifier applications; and (3) photosensitive glasses which can be used both for direct writing of channel waveguides as well as writing of Bragg gratings. Different sol-gel techniques have been employed both for bulk, as well as waveguide applications. The sol-gel process has also been used to introduce organic molecules and enzymes in gels to fabricate various sensors.

One of the difficulties when it comes to the fabrication of thick sol-gel films required in integrated optics is that the film cracks if the deposited layer is thicker than a certain value, sol-gel process based on the repetitive spin coating and rapid thermal annealing which overcomes this problem has been developed at Imperial College, London, U.K. The method use a multilayer processing in which thermal annealing is done at the proper temperature after the application of each individual layer by spin coating [29]. A dip-coating method along similar lines has also been reported, A reflow and buried technique to form a sol-gel glass channel waveguide on silicon has also been studied. Sol-Gel syntheis of fluoride optical materials has been studied for IO application another important method involves the use of organically modified silicat system (ORMOSIL). This hybrid system used organo-minera precursor and the Sol-Gel process. This way, one can obtain UV ligh to directly pattern channel waveguides or other devices bypassing the need for a mask and etching procedure.

Sol-Gel waveguide doped with rare-earth elements have been studied by orignac et al. [30]. A also vaule of ~1 dB/cm and a fluorscence life time of 375 ms were measured for the case of Nd-Doped Sol-Gel film, while for the case or Er-doped film, the fluorescence life time was 1.78 ms.

6.9 Fabricating Microring Amplifier on the Same Substrate

Referring to the scheme of Erbium doped fiber amplification as described in First Chapter of the thesis, we will require to couple the pump wavelength λ_p and selected wavelength λ_3 through a wavelength selective coupler into the Erbium doped resonating Microring and then it has to be coupled to output waveguide as amplified signal. This all arrangement can be made on a single chip using the principles of Silicon Optical Bench (SiOB) waveguide technology [31] and combining the lateral and vertical coupling system as explained in section 5 of this chapter.

A seemingly trivial hurdle has to be taken to arrive at these small waveguide structures, light should be transported without losses through bends with a radius of only a few micrometers. This can only be achieved by careful design and working with materials allowing high index-contrast structures. High index-contrast waveguides used for microring resonators support tightly confined modes, that is, modes whose energy is contained almost entirely within the waveguide core. There are principally two arrangements for the positioning of the adjacent waveguides with respect to the resonator, the lateral and vertical coupling. The vertical arrangement requires a two step lithographic process. Here the coupling constants are mainly determined by the thickness and refractive index of the intermediate layer and the relative offsets of the underlying waveguides with respect to the ring.

This approach allows for an optimised independent choice for ring and port waveguides. Critical in the vertical arrangement is the alignment of the two lithographic steps, where a precision within 100nm is needed. In the case of lateral coupling only a single lithographic step with a single mask in needed. The coupling is mainly determined by the width of the gap between the straight and bent waveguides and demands nanometer precision in the case of high refractive index contrast.

In Silicon Optical Bench Technology optical waveguide circuits are made from doped silica films deposited on planer substrate. The base layer is made of undoped silica which is the most rigid layer and it keeps the input pump signal (λ_p) waveguide and the amplified output signal wave guide at a spacing. Grown over this layer is another layer to contain erbium doped microring and a laterally coupled waveguide to carry input signal that is to be amplified. Here the distance and layer thickness is decided to achieve vertical coupling of pump wavelength lp and output wavelength. In this arrangement we could dispense with wavelength selective coupler and isolators by using the principle of coupled microring resonator thus achieving compactness and economy.

The pump signal is applied in the base layer waveguide whose distance with the microring is arranged so that pump wavelength λ_p gets vertically coupled to erbium doped microring resonator and create population inversion in the ring structure. Now when

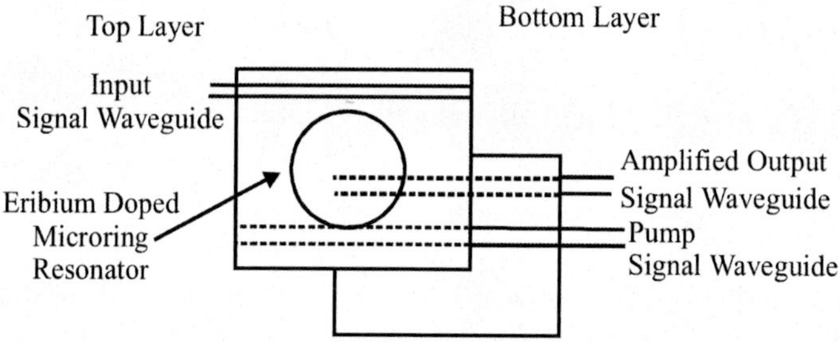

Fig 6.6 Schematic of Micro Ring Amplifier

the input signal to be amplified passes through top layer then required wavelength λ_3 gets coupled to the resonator where stimulated emission and rounding takes place. Amplified wavelength is arranged to get vertically coupled to output waveguide in the base layer which is designed to collect more power than the input waveguide.

6.10 Conclusion

In this work the fabrication of integrated optical waveguides and microring resonators with dopants like erbium, neodymium and cerium is studied. It is found that sol-gel process is an effective method of fabricating microring amplifiers. In recent years there has been rapid progress in the fabrication and characterisation of sol-gel based IO waveguide and devices, with and without active dopant. However till date. I have not found any report claiming packaged device ready for commercial exploitation. Therefore the flexibility and economy offered by sol-gel technology is yet to become practical in communication application.

6.11 Further Scope of Work

In this Chapter a Novel Microring Amplification technique is proposed, discussed and analysed. The Lay out of complete amplifier is prepared and ready for fabrication. The microring are gaining popularity as photonic is tending towards integrated photonic bypassing integrated opto-electronics. With the emergency of nano-technology and nano photonics, the discussed technology in this work proves to be a viable approach.

Further work in this area, can be pursued in exploring the fundamental properties of microring resonators based on rigorous wave theory. This is a challenging task since many aspects are yet to be explored such as wave theory of curved waveguides, coupled mode theory for non-uniform structures etc. This also enables analysis of ring resonators of much smaller dimensions, such as those with circumference of the order of wavelength of light. Presently, many non-uniform structures are analysed using numerical techniques but they do not provide much insight into the actual phenomena.

References

[1] F. Morichetti, A. Melloni, and M. Martinelli, "Effects of Polarization Rotation in Optical Ring-Resonator-Based Devices," Journal of Lightwave Technology, vol. 24, No. 1, pp. 573-585, Jan. 2007.

[2] K. Greene. (2007) Intel speeds up silicon photonics. Internet draft. [Online]. Available: http://www.technologyreview.com/lnfotech/18087/

[3] (2007) World's Fastest Optical Chip. Internet draft. [Online]. Available: http://www.technologyreview.com/lnfotech/18000/page1/

[4] K. Bullis. (2007) A Nano Solution to Increasing Bandwidth. Internet draft. [Online]. Available: http://www.technologyreview.com/lnfotech/18074/

[5] A.D.Boardman and P.Egan, "S-Polarized Waves in a Thin Dielectric Film Asymmetrically Bounded by Optically Nonlinear Media," Journal of Quantum Electronics, vol. QE-21, no. 10, pp. 1701-1713, Oct. 1985.

[6] G. I. Stegeman and C. T. Seaton, "Nonlinear Integrated Optics," Journal of Applied Physics, vol. 58, no. 12, pp. R57-R78, Dec. 1985.

[7] B. E. Little and S. T. Chu, ''Towards Very Large-Scale Integrated Photonics," Optics & Photonics News, pp. 24-28, Nov. 2000.

[8] W. Krolikowki and Y. S. Kivshar, "Soliton-based optical switching in waveguide arrays," J. Opt. Soc. Am. B, vol. 13, no. 5, pp. 876-887, May 1997.

[9] D. W. Vernooy, A. Furusawa, N. P. Georgiades, V. S. llchenko, and H. J. Kimble, "Cavity QED with high-Q whispering gallery modes," Physical Review A, vol. 57, No. 4, pp. R2293-R2296, Apr. 1998.

[10] Sandoghdar, F. Treussart, J. Hare, V. Lefe'vre-Seguin, J.-M. Raimond, and S. Haroche, ''Very low threshold whispering-gallery-mode microsphere laser," Physical Review A, vol. 54, no. 3, pp. RI777-RI780, Sept. 1997.

[11] V. B. Taranenko and C. O. Weiss, "Spatial Solitons in Semiconductor Microresonators," IEEE Journal of Selected Topics in Quantum Electronics, vol. 8, no. 3, pp. 488-496, May/June 2002.

[12] O. Schwelb and 1. Frigyes, "All-Optical Tunable Filters Built with Discontinuity Assisted Ring Resonators," Journal of Lightwave Technology, vol. 19, no. 3, pp. 380-386, Mar. 2001.

[13] M. Kohtoku, S. Oku, Y. Kadota, Y. Shibata, and Y. Yoshikuni, "200-GHz FSR Periodic Multi/Demultiplexer with Flattened Transmission and Rejection Band by Using a Mach Zehnder Interferometer with a Ring Resonator," IEEE Photonics Technology Letters, vol. 12, no. 9, pp. 1174-1176, Sept. 2000.

[14]. T. A. Ibrahim, W. Coo, Y. Kim, J. Li, J. Goldhar, P.-T. Ho, and C. H. Lee, "All-Optical Switching in a Laterally Coupled Microring Resonator by Carrier Injection," IEEE Photonics Technology Letters, vol. 15, no. 1, pp. 36-38, Jan. 2003.

[15] H. Takahashi, R. Inohara, M. Hattòri, K. Nishimura, and M. Usami, "Expansion of compensation bandwidth of tunable dispersion compensator based on ring resonators," Electronics Letters, vol. 40, no. 16, pp. 1014-1016, Aug. 2004.

[16] J. Azaa and L. R. Chen, "Multi wavelength Optical Signal Processing Using Multistage Ring Resonators," *Photonic Tech., letters,* vol. 14, no. 5, pp. 654-656, May 2002.

[17] S. T. Chu, B. E. Little"W. Pan, T. Kaneko, and Y. Kokubun, "Cascasded Microring Resonators for Crosstalk Reduction and Spectrum Cleanup in Add-Drop Filters," IEEE Photonic Technology Letters, vol. 11, no. 11, pp. 1423-1428, Nov. 1999.

[18] R. Grover, T. A. Ibrahim, S. Kanakaraju, L. Lucas, L. C. Calhoun, and P.-T. Ho, "A Tunable GaInAsP InP Optical Microring Notch Filter," IEEE Photonics Technology Letters, vol. 16, no. 2, pp. 467-469, Feb. 2004.

[19] L. Martinez and M. Lipson, "High confinement suspended micro-ring resonators in silicon-on-insulator," Optics Express, vol. 14, no. 13, pp. 6259-6263, June 2007.

[20] S. Gulde, "Optimization of ultrafast all-optical resonator switching," Optics Express, vol. 13, no. 23, pp. 9502-9515, Nov. 2005.

[21] G. Paloczi, Y. Huang, and A. Yariv, "Free-standing all-polymer microring resonator optical filter," Electronic Letters, vol. 39, no. 23, pp. 90-103, Nov. 2003.

[22] H. Ou, K. Rottwitt, and H. Philipp, "Deep glass etched microring resonators based on silica-on-silicon technology," *Electronics Letters,* vol. 42, no. 10, May 2007.

[23] A. Melloni, R. Costa, P. Monguzzi, and M. Martinelli, "Ring-resonator filters in silicon oxynitride technology for dense wavelength-division multiplexing systems," *Optics Letters,* vol. 28, no. 17, pp. 1567-1569, Sept. 2003.

[24] A. M. Armani and K. J. Vahala, "Heavy water detection using ultra-high-Q micro cavities," *Optics Letters,* vol. 31, no. 12, pp. 1896-1898, 2006

[25] H. P. Weber and R. Ulrich, "A thin-film ring laser," *Applied Physics Letters,* vol. 19, No. 2, pp. 38-40, July 1971.

[26] J. Haavisto and G. A. Pajer, "Resonance effects in low-loss ring waveguides," *Optics Letters,* vol. 5, no. 12, pp. 510-512, Dec. 1980.

[27] J.Y. Allain, M. Moneric; "Room temperature CW tunable green upconversion holmium fiber laser"; Electron Letters, Vol. 26, pp 261-263, 1900

[28] SR Natarajan; "Theoretical and Experimintal Studies on Optical amplification in sol-gel films"; Ph.D. Dissertation, IIc. Bangalore India, April 2000

[29] A.S. Holmes and M green; "Fabrication of burried channel waveguides on silicon sabstrates using spin-on glass"; J. App optics' Vol. 32, No. 25, pp. 4916-4921, 1993

[30] X. Orginac et all; "Fabrication and characterisation of sol-gel planer waveguides doped with rare earth ions"; Applied Physics letter, Vol.-69, No. 12, pp. 895-897, 1996

[31] Yaan P. Li & Charles H. Henry; "Silicon Optical Bench Waveguide Technology"; Optical Fiber Telecommunication, Vol. III B, 2000, pp (319-375)